Touch

Touch

Gabriel Josipovici

Yale University Press
New Haven and London

For permission to reprint extracts from copyright material the author and publishers gratefully acknowledge the following: Faber and Faber Ltd for 'I'm cross with god' (Song 153) from *The Dream Songs* by John Berryman and for *New Year Letter* by W. H. Auden; Random House Inc. for *New Year Letter* by W. H. Auden; Princeton University Press for *The Divine Comedy* by Dante translated and edited by C. S. Singleton; Penguin Books for *The Theban Plays* by Sophocles translated by E. F. Watling.

Set in Garamond Simoncini by SX Composing
Printed in Great Britain by St Edmundsbury Press

Library of Congress Cataloging-in-Publication Data

Josipovici, Gabriel, 1940–
    Touch / Gabriel Josipovici.
    Includes index.
    ISBN 0–300–06690–2 (alk. paper)
    1. Nonverbal communication. 2. Touch. I. Title.
P99.5.J67 1996
302.2'22—dc20   96–19883
                                                                                    CIP

A catalogue record for this book is available from the British Library.

10 9 8 7 6 5 4 3 2 1

For Margreta, who will know why.

Put your finger into every bottle, to feel whether it be full,
which is the surest way, for feeling hath no fellow.

Jonathan Swift, *Directions to Servants*

Contents

Contents

Acknowledgements

Though this is a very personal book it grew in large part out of conversations and debates with friends. An animated discussion on the subject of touch with Stuart Hood was followed by a long letter from him, parts of which were so illuminating that I have quoted them in the Appendix. Along the way I showed versions of the book to many people, who all made helpful suggestions: Monika Beisner, Rosalind Belben, Brian Cummings, Paul Davies, Dan Gunn, Sue Loewenstein, John Mepham, Desa Philippi, Sacha Rabinovitch. Bernard Harrison and Margreta de Grazia read both an early and a late draft with a care far beyond the call of duty or even friendship, covering every page of typescript with helpful criticism and comment. The book would not be what it is without them, even though in the end I am afraid I often ignored their advice and went my own way. Robert Baldock and Candida Brazil, my editors at Yale, could not have been more helpful. To all, my grateful thanks.

Prologue

The desire to write an essay on touch has grown steadily on me in the last few years.

I do not know what will be in that essay or how I will go about writing it, but the thought of writing it will not leave me.

For some time I have been jotting down notes to myself on the topic of touch, collecting postcards and quotations that seem relevant, and trying out my ideas on my friends. But the gap between this preliminary work and the actual writing of the essay is unbridgeable. Once I have written down the sentence, 'The desire to write an essay on touch has grown steadily on me in the last few years', I have entered a new world, a world where what I had previously thought and read and jotted down is about as useful as those conjectural maps of an unexplored region are for the explorer: not entirely useless, but of little immediate practical help. How I advance from now on, and if I advance at all, is going to depend less on those thoughts and jottings than on the resources I can call on daily within myself and on the decisions I take as I go along.

It is important, in such circumstances, to feel one's way forward with care, but also never to stand still for too long.

These are of course merely metaphors. Why do they seem to correspond to my (as yet unclear but not exactly vague) aims?

1

It is, I think, because the notion of feeling one's way forward, of groping in the dark or semi-darkness, implies a testing of the way with the whole body. And although this method may be painfully slow, it is much less likely to lead me astray than if I relied on sight alone and had open country to cross and a bright sun to go by. In this way I will experience every inch of the way rather than suddenly finding that I have reached my goal with very little sense of the terrain I have passed through. If I can simply walk across the space that lies between me and my goal I may arrive there quickly, but then I will be left wondering whether I have really arrived or only dreamed or imagined it.

The ground I have to cover does not lie stretched out before me, but it does not lie inside me either. Where then does it lie? Or do I need another metaphor altogether? This is one of the things I will no doubt discover in the course of writing this essay. All I know at the moment is that my instinct tells me that I have to rely on touch rather than on sight, on groping my way forward rather than striding or running. But that, surely, is appropriate to the subject in hand.

In hand? That is surely the point. I do not have the subject in my hand. I do not hold it. But where then is it? And do I have to try and banish all such metaphors as misleading as I set out in search of it?

I don't think so, for then I would not be able to write at all. Perhaps all I need to do is note our language's propensity for such metaphors and see if that tells me something about the road I have to follow.

Enough. It is time to begin.

1 The Lesson of the Hand

There is a scene in Charlie Chaplin's *City Lights* which, when I first saw it, stirred me as I am rarely stirred, and has remained firmly etched in my mind in all the intervening years (though I realise, as I come to write about it here, that my memory of it contains many blanks and that my clear visual recollection is perhaps more a recollection of the *effect* of that scene than of its precise details – but let that be, I have no wish to see it again in order to check, for what interests me is to understand why that first experience was so powerful and why I have never forgotten it).

That mildly funny and heavily sentimental film deals with a subject to which drama and film have occasionally been drawn: blindness. It concerns the relations between a little tramp and a beautiful blind flower-girl. I do not recall anything about the plot, only its climactic scene. The little tramp has been kind to the girl, has even, I seem to remember, managed to procure money with which she can have an operation to restore her sight. The operation is successful, but she does not know who her benefactor is, for the tramp has kept his identity hidden, though managing to convey the impression that he is a very rich man. We now see the girl, her sight restored, working in a smart flower-shop and chatting to her fellow-employee. We see her through the plate-glass window of the shop, through the serried ranks of flowers. We see her, in fact, through the eyes of the little tramp, who is passing by, glances in, then stops in astonishment as his eyes light on her.

She, looking out into the street as she talks and laughs with her friend, gradually becomes aware of him. Through the shop window we see her glancing at him and then laughing, perhaps at his comic appearance, perhaps just at something the other girl has said. However, as he stands on the pavement looking in (and in my memory at least the camera stays resolutely with him throughout), the two girls gradually become aware that something is amiss. Even a hungry and exhausted tramp would not go on staring in as this man is doing. What does he want? Why doesn't he move on? The girl and her friend talk seriously for a moment, then she takes a flower from a bucket, opens the door of the shop and steps outside. She holds the flower out to him: Here. For me? Yes, take it. No, I couldn't, really. But you must. No no. Please take it. (It is important for all this that it is a silent film – the last silent film, I seem to recall, that Chaplin made.) Finally he accepts the flower. She turns and steps back inside the shop.

For the audience the scene has become unbearable. The girl is there, so close to her benefactor (who could have left the pain and sorrow of a tramp's life behind if he had not selflessly given the money that fell into his hands to help her recover her sight), and yet she does not and cannot know him. (Even were he capable of it, it would be no use his telling her of his role in her life, since she would not believe him.) Will that be how it remains? Will that be *it*?

In such circumstances, as I sit in the dark watching the film unfold, my own response is silently to beg the scriptwriter and director to have pity on me, not to harden their hearts but to allow the longed-for recognition to take place. This, I realise, is an odd response. I could understand an unsophisticated viewer praying to whatever god he believed in to make the recognition happen; and I could understand a sophisticated viewer refusing to be taken in by what is, after all, only a popular film, and so shrugging the whole thing off as a pathetic attempt to pull the audience's heart-strings, a crude ploy from which he will resolutely withhold his assent. What I cannot understand is my simultaneous awareness of the fact that this is only a film, and a rather meretricious one at that, *and* my desperate wish that whatever human being is manipulating the strings should have pity on my suffering and not allow the tramp and the flower-girl to part without her recognising him.

The little tramp has remained on the pavement, still looking into the shop where the girl is once again to be seen talking and laughing with her friend. His presence is clearly an embarrassment to the girls, for they keep glancing out in his direction and now are obviously arguing about what to do about him. Finally the girl turns to the till, opens it and takes out some coins. She steps out into the street once more, and once more approaches the little tramp. Now she is trying to get him to accept the money, but he keeps refusing: No no, I wouldn't dream of it. Please. No no. Please, you must. No I couldn't, really.

Think how such a scene would be in a talkie. How embarrassing the dialogue. How soon it would be over. But scenes such as this are the stuff of the silent cinema, as he mimes dignified refusal and she mimes passionate entreaty. Finally, as we had prayed and hoped – but would it ever happen? – she takes his hand and thrusts the coins into it.

In my memory the camera has settled on her face, and what follows is seen as he sees it, close to. She does not need to be a particularly good actress. Indeed, it would be a mistake to try and get her to register in detail on her face what is happening to her. For, in a strange way, it is now we who are doing all the work. It is we who hold his hand in ours. It is we who suddenly become aware of that hand as somehow familiar. It is we who, in a long instant, make the amazing, the impossible connection.

That instant is not just one of the most powerful filmic experiences I have ever had, it seems to transcend art altogether. It is what Barthes, trying to understand why certain photos had the power to move him while others left him cold, however much he might admire them, called a *punctum*. Barthes had to elaborate a complex theory about photographs to account for this, suggesting that the essence of photography is the message: X was *there* (and where is he/she now?). The explanation of why an improbable and sentimental recognition scene such as this one seems to tear itself right out of the fabric of the film in which it is embedded and strike at the very core of my being seems even more puzzling. For what have the flower-girl and the little tramp to do with me? An hour and a half ago I did not even know of their existence. Since then I have watched their antics on the screen, laughed occasionally, been occasionally moved, but nothing had led me to expect this feeling of being opened up, simultaneously destroyed and reconstituted, which I am now experiencing. What is going on?

We think (when we stop to think about it at all) of other people as occupying an objective space in front of us, and of our knowledge of them as being derived from our ability to see. But this is not in fact how we apprehend others. At least part of what enters into an apprehension of them is our common bodily and kinaesthetic reaction to a physical world which we both inhabit. For we are embodied, and it is our bodies which give us common access to the physical world; in other words we are participators, not spectators, and it is through embodiment that we participate. Merleau-Ponty, who has done more than anyone else to draw this fact to our attention and to draw the consequences from it, illustrates this with the following anecdote:

> I am watching this man who is motionless in sleep and suddenly he wakes. He opens his eyes. He makes a move toward his hat, which has fallen beside him, and picks it up to protect himself from the sun. What finally convinces me that my sun is the same as his, that he sees and feels it as I do, and that after all there are two of us perceiving the world, is precisely that which, at first, prevented me from conceiving the other – namely that his body belongs among my objects, that it is one of them, that it appears in my world. When the man asleep in the midst of my objects begins to make gestures toward them, to make use of them, I cannot doubt for a moment that the world to which he is orientated [que le monde auquel il s'adresse] is truly the same world that I perceive.

Why do I immediately and intuitively understand what is going on? The answer, perhaps surprisingly, is because the man is neither an object in my visual field nor myself. If he were merely an object in my visual field I would have to try and grasp the meaning of his gestures by an elaborate system of interpretation that would remain for ever incomplete. This is the situation of the protagonist in the novels of Kafka and Robbe-Grillet, but we feel that it is unusual, counter-intuitive. The protagonists of those novels seem to bear the burden of *having to understand* rather than simply living, and the novels show that this is an unnatural state to be in, even if it is one the protagonists seem condemned to endure. But even if the man was, somehow, me (if, without my having realised it, I was, for example, seeing myself in the mirror), then exactly the same thing would occur, as I ran round my subjective labyrinth seeking an explanation.

This is what happens, for example, in Beckett's *Molloy*, where Molloy, looking at his own hand or leg, cannot make much sense of it but comes to the tentative conclusion that it must be his since it is hardly likely to belong to anyone else. But again, while we have no difficulty in understanding this (it is the kind of thing that happens to us from time to time), we feel that Molloy's relation to his own body is unnatural (and are compelled to laugh at it). But, as Merleau-Ponty notes, the man in his example is there 'in a certain place in my [visual] field, but that place at least was ready for him ever since I began to perceive'. It is 'the experience that I make out of my hold on the world' which 'makes me capable of perceiving another myself, provided that in the interior of my world there opens up a gesture resembling his own'.

It is because I move in the same world as other people that I can respond to them as to myself. It is because the man I see is asleep under the same sun as I who see him live under, because he is asleep amid objects I am myself familiar with, in a landscape I too have rested in, with a hat such as I too might have worn, and making a gesture I myself have often made, that I have no doubts at all about what it is he is doing and why he is doing it. No particular empathy is required, merely the innate knowledge of my own body as existing in the world.

But if that is the case then the phenomenon of film is profoundly paradoxical. The world appears to us on a screen in a darkened room and with no effort at all we enter it. We inhabit that world as we inhabit our own, but there is one crucial difference: *we have left our bodies behind*. The advantage of this is that we can live out vicariously all sorts of adventures, knowing in the core of our being that we will always be safe. The disadvantage is that, having left our bodies behind on the seat of the cinema, nothing that happens on the screen can truly affect us.

Film thus allows us to come alive for a time, to act and suffer in ways far more decisive and meaningful than we can ever experience in ordinary life, and to repeat that whenever we want (if we have the means to satisfy that want). But though it feeds a genuine human hunger, its food is of poor nutritional value. For the sky under which the man is waking up, on film, is not the same as the one under which I am watching him. He exists so palpably there before me that I can wake up with him, shield myself from the sun with him, *but I cannot cross over to where he is lying*. I know that, but the paradox is that I can only experience him if I deny that knowledge, and yet my experience seems to be predicated on that ability to cross over and nudge him. There is thus a profound contradiction in my experience of him, and the repression of that contradiction from my consciousness, allied to my hunger for his presence, contains, as we will see, all the ingredients necessary to addiction.

Yet what film normally withholds from us, that scene in *City Lights* offers us. For if the girl can now see, and is thus no different from the rest of the characters in the film and from the audience in the darkened room, it is nevertheless her past blindness that is in play here. As she takes the little tramp's hand to press the coins upon him we watch her discover that she already knows that hand, and knows it in a way quite different from the way we, who have been seeing it on screen for an hour or more, can be said to know it. And yet, because we have lived her blindness with her, we too become privy to a discovery being made at a far deeper level than the other transactions on the screen, a discovery that is being made in the secret recesses of her body. Secret of course not because they are hidden from sight (which would suggest that they might one day be revealed), but because it is *her* body, existing in time, with its own memories and its own life. We discover that, as I have said, not because we can read her face but because we too have bodies and we too know what it is to experience with our bodies rather than merely to see.

The encounter thus gives back, to the girl and to us, that inwardness with the body which the simple experience of witnessing people and events, whether on screen or in real life, denies us. By depicting, for my sight, for once, that which takes place where sight has no jurisdiction, the scene does nothing less than give me back a sense of my own body, not as an object but as that which is alive in space and time.

2 The Lime-Tree Bower and the Virgin of Amiens

Sight is free and sight is irresponsible. I can cast my eye to the far horizon and then back to the fingers I hold up before my face, all in a fraction of a second and with no effort at all. And I can repeat the operation at will. On the other hand, were I to walk to that point on the horizon it would take time and effort, time and effort which I might feel I could better employ doing something else. To look costs me nothing but to go involves both a choice and a cost.

Yet the very decision to walk to that place endows the ensuing walk with a weight which mere looking lacks. A walk will always bring me up against the unexpected: a new view of the place I have just left; a tree I had never seen before; an encounter with a friend; an accident. None of that would have been experienced by me if I had been content to look from the safety of my window. But it is not that experience in itself is necessarily valuable. It is that the letting go of oneself which is involved in any sort of setting out, whether for a three-year journey or a ten-minute walk, is also, mysteriously, a bringing to life of parts of oneself which had previously lain dormant.

The uncanny nature of both photography and film has something to do with the way that both of them reinforce the distinction between sight and bodily experience, yet keep that distinction hidden. 'Photography maintains the presentness of the world by accepting our absence from it', Stanley Cavell has written in his perceptive book on film. 'In viewing a movie', he remarks a little later, 'my helplessness is mechanically assured: I am present not at something happening, which I must confirm, but at something which has happened.'

Of course physical displacement is not precisely the issue. Drawing on Cavell's insights we might say that the difference between two kinds of looking lies in our sense of being either present or absent-and-merely-looking. Walter Benjamin coined the term 'aura' to describe the first, specifically in the context of his exploration of the nature of film. 'To perceive the aura of an object we look at', he writes, 'means to invest it with the ability to look at us in return.' This reciprocity, he suggests, depends on our sense of the uniqueness of the moment; aura is 'the unique phenomenon of a distance, however close it may be. If, while resting on a summer afternoon, you follow with your eyes a mountain range on the horizon or a branch which casts its shadow over you, you experience the aura of those mountains, of that branch.' Aura does not abolish distance, to adapt a wonderful phrase Benjamin once used about friendship, it brings distance to life. In so doing it pierces us with the sense of the uniqueness, the unrepeatableness, of the occasion. Aura is what envelops an object as we experience wonder not at this or that aspect of it but at the simple fact that it is, and that we, observing it, are. Mechanical reproduction, whether through photography or film, denies and destroys aura. The peculiar charge of that climactic scene in *City Lights* lies in the momentary appearance of aura in a medium in which aura has no place.

The loss of aura is what Coleridge mourns in his great Dejection Ode:

> I see them all so excellently fair,
> I see, not feel how beautiful they are!

The loss of aura means the loss of presence: somehow, I am not here to respond to this, even though I can monitor (the technical term seems appropriate) my inability to respond. And the loss of presence means in some sense the loss of the body. We exist but, because we cannot feel, because the world seems no longer to be looking at us, our existence is devoid of meaning and hope. We can go about our daily lives as we have always done, but we might as well be prisoners in solitary confinement.

The Lime-Tree Bower and the Virgin of Amiens

As it happens, Coleridge wrote another dejection ode, 'This Lime-Tree Bower My Prison', which explores all these interconnected themes: how walking brings the body to life, the sense of loss of self which imprisonment produces, and the nature of aura. 'In the June of 1797', he writes in the note he placed at the head of the poem,

> some long-expected friends paid a visit to the author's cottage; and on the morning of their arrival, he met with an accident, which disabled him from walking during the whole time of their stay. One evening, when they had left him for a few hours, he composed the following lines in the garden-bower.

The ensuing poem is a typical Romantic ode in that it conveys the sense of discoveries being made in the actual course of the writing. Its progress must therefore be followed with some care.

It begins with an elaboration of the sentiment contained in the head-note:

> Well, they are gone, and here must I remain,
> This lime-tree bower my prison! I have lost
> Beauties and feelings, such as would have been
> Most sweet to my remembrance even when age
> Had dimmed mine eyes to blindness! They, meanwhile,
> Friends, whom I never more may meet again,
> On springy heath, along the hill-top edge,
> Wander in gladness . . .

As he writes though, and as he imagines them on their walk, he seems to forget his confinement and to be out there with them:

> They, meanwhile . . .
> Wander in gladness, and wind down, perchance,
> To that still roaring dell, of which I told;
> The roaring dell, o'erwooded, narrow, deep,
> And only speckled by the mid-day sun;
> Where its slim trunk the ash from rock to rock
> Flings arching like a bridge; – that branchless ash,
> Unsunned and damp, whose few poor yellow leaves

Ne'er tremble in the gale, yet tremble still,
Fanned by the water-fall! and there my friends
Behold the dark green file of long lank weeds,
That all at once (a most fantastic sight!)
Still nod and drip beneath the dripping edge
Of the blue clay-stone.

It is very difficult to know where to stop quoting with these Romantic poems, for firm initial statements ('they are gone,' 'I have lost') seem to open up and develop into a rhythmic continuum where certain key words act like stepping-stones, leading one ever forward. As Donald Davie says of this passage:

When the poet thought all had been said, it turned out that nothing had been said; in calling to mind the pleasures he cannot share, his imagination permits him to share them. This refers back to the paradox which gives the poem its title. How can a bower of lime-trees be a prison? And, even as he begins to show how this can be, he proves that it cannot be, since the imagination cannot be imprisoned.

Yet I think Davie goes a little too fast here. It is not the imagination that releases Coleridge, but the act of writing. 'Dell', 'ash' and 'water-fall' do not form links in a chain of imagining but of writing. The movement of the poem under his hand is what releases the imagination, as the act of walking would have given him that which he would have been able to recall even in old age.

When, in the second stanza, though, there was the danger that the poem might peter out, overcome by its effort to keep up with the rest of the group, so to speak, it is renewed not by the poet's determination to go on imagining that walk but by the sudden rising within himself, as he writes, of the peculiar joy the walk must be giving his city friend, Charles Lamb:

for thou hast pined
And hungered after Nature, many a year,
In the great City pent, winning thy way
With sad yet patient soul, through evil and pain
And strange calamity!

As in the Dejection Ode it is the thought of another which lifts the poet out of the prison of his dejection, so here first the thought of the whole group and then of his beloved friend and the joy he must be taking in his walk arouses the imprisoned poet to a totally unexpected happiness. And it is this which leads to the surprise of the final stanza:

> A delight
> Comes sudden on my heart, and I am glad
> As I myself were there! Nor in this bower,
> This little lime-tree bower, have I not marked
> Much that has soothed me. Pale beneath the blaze
> Hung the transparent foliage; and I watched
> Some broad and sunny leaf, and loved to see
> The shadow of the leaf and stem above
> Dappling its sunshine! And that walnut-tree
> Was richly tinged, and a deep radiance lay
> Full on the ancient ivy, which usurps
> Those fronting elms, and now, with blackest mass
> Makes their dark branches gleam a lighter hue
> Through the late twilight; and though now the bat
> Wheels silent by, and not a swallow twitters,
> Yet still the solitary humble-bee
> Sings in the bean-flower!

What Coleridge has in fact discovered in the course of writing the poem is that the lime-tree bower is not a prison, not because he can escape, either in reality or in imagination, but because, if he will only look around him, he will see that there is much to wonder at, much that he had never noticed, as much here to feed his memory in his old age as in the most protracted walk.

As so often though in the odes of Coleridge and Wordsworth, the poet himself moves too quickly from the experience and from the detail of what he has just written to some general and uplifting conclusion:

> Henceforth I shall know
> That Nature ne'er deserts the wise and pure;
> No plot so narrow, be but Nature there,
> No waste so vacant, but may well employ

> Each faculty of sense, and keep the heart
> Awake to Love and Beauty! and sometimes
> 'Tis well to be bereft of promised good,
> That we may lift the soul, and contemplate
> With lively joy the joys we cannot share.

This is too pat. The poem has not been about Love or Beauty or Heart or Soul. It has not been about Christian resignation or Stoic resourcefulness. What it has been about is the ebb and flow of feeling, about the interaction between self and place, about the failure of feeling and the return of feeling, as memory, imagination and the world around one set poetry going, and poetry, in turn, sets the imagination working, and imagination is rewarded and rewards.

Aura, said Benjamin, is dependent on a sense of reciprocity: 'To perceive the aura of an object we look at means to invest it with the ability to look at us in return.' That is what the first part of the last stanza shows us: 'Pale beneath the blaze/ Hung the transparent foliage; and I watched/ Some broad and sunny leaf, and loved to see/ The shadow of the leaf and stem above/ Dappling its sunshine!' Dappling here, as in Hopkins, though far less consciously, is the sign of life itself. It is, if we think about it, a word which describes something we see in terms of our not exactly seeing an object and not exactly having a mere impression. The play of light and shade is precisely that, play, movement, as the leaves move in the tree and the light of the sun is broken and constantly altered. And as we read these lines and note the rapt precision of Coleridge's description, Benjamin's other remark about aura comes to mind: that aura is 'the unique phenomenon of a distance, however close it may be'. Coleridge sits in his little bower and as he writes he moves towards a sense of the miraculous nature of *now*, of what he sees now and of the distance that separates him from and binds him to tree, leaves, shadow, sunlight, breeze: 'I watched/ . . . and loved to see/ . . . And that walnut tree/ Was richly tinged, . . . and *now*, with blackest mass/ . . . and though *now* the bat/ Wheels silent by, and not a swallow twitters,/ Yet still the solitary humble-bee/ Sings in the bean-flower!'

I am reminded, as I think about aura and Coleridge's poem, of the distinction Proust makes in an early essay between his response to a statue of the Virgin on the south porch of Amiens Cathedral and the *Mona Lisa* in the Louvre: 'I feel I was wrong to call it a work of art', he says of the Virgin,

> a statue that is . . . forever part of a particular place on earth, of a certain city, that is to say a thing which has a name as a person has, which is an individual, of which no other exactly alike can ever be found on the face of the continents, of which even the railway employees at the place where we must inevitably come in order to see it, in announcing its name seem unwittingly to tell us: 'Love that which you will never see twice' – such a statue perhaps has something less universal than a work of art; it holds us, at any rate, by a tie stronger than that of the work of art itself, one of those ties such as persons and countries hold us by.

The *Mona Lisa*, though obviously painted by a certain person in a certain place, is now uprooted, rather 'like a wonderful woman "without a country" [quelque chose comme une admirable "sans-patrie"]', though, since she is in the Louvre she is in a sense a naturalised French subject. Think, on the other hand, he goes on, of her sister, the Virgin of Amiens, 'smiling and sculptured':

> Coming no doubt from the quarries near Amiens, having made only the one trip in her youth to come to the porch of Saint Honoré, not having moved since, having been weather-beaten little by little by the damp wind of the Venice of the North which bent the spire above her, gazing for so many centuries on the inhabitants of this city of whom she is the oldest and most sedentary, she is truly an Amiénoise. She is not a work of art. She is a beautiful friend we must leave at the melancholy provincial square from which no one has ever succeeded in taking her away, and where, for eyes other than ours, she will continue to receive directly on her face the wind and sun of Amiens and to let the little sparrows alight with a sure instinct . . . in the hollow of her welcoming hand.

And Proust ends this contrast with one of those simple but devastating remarks which lift even these early essays into the realm of greatness: 'In my room a photograph of the *Mona Lisa* retains only the beauty of a masterpiece. Near her a photograph of the Vierge Dorée takes on the sadness of a souvenir [prend la mélancolie d'un souvenir].'

That one word 'only' – 'seulement' – consigns a century of aesthetics to the dustbin. For what Proust is saying here, as though it were the most natural thing in the world, is that in the end aesthetic beauty, the notion of the masterpiece, is completely trivial. The Virgin of Amiens may be inferior artistically to the *Mona Lisa* (but would Proust have been so categorical today?), but it affects us far more profoundly than any merely aesthetic experience ever could, striking at the core of our being, like those involuntary memories *A la recherche* will go on to evoke with such eloquence and precision.

The power of memory for Proust lies in the fact that it brings back to us what we had lost so thoroughly that we were not even aware of having lost it. Marcel has mourned his grandmother, of course, but it is only when his body inadvertently repeats a movement it had made in her presence that her death is borne in upon him – in the very instant that he experiences her alive in a way she had, ironically, never been for him when she was actually alive, the magnitude of the loss that is her death hits him for the first time. The experience makes him understand that we are creatures who exist in time, who were once what we are no more, who once had what we can never have again. Our visit to Amiens is precious to us, and the Virgin forms part of that visit, like the colour of the sky, the quality of the breeze, the smell of the train that brought us there, for we experience, in that sleepy provincial town, something we had not been prepared for and which we will find nowhere else, not even there, were we to return. It is not nostalgia which overwhelms Proust but the sense of himself brought to sudden life by the encounter with the cathedral and its sculptures, the sense of possibilities released in him by the encounter which he did not know were there. This sensation, for Proust as for Coleridge, is instinctively linked with joy, and even a cheap postcard reproduction of the Virgin will be enough to remind him of what he owes her, of what she means to him.

It is not of course that a visit to the Louvre cannot lead to a similar experience, but the wrenching of works of art from their matrix and the bundling of so many disparate works into one building militates against this. Of course the Virgin of Amiens was made for a particular place on a particular building, and centuries of sun and rain have ensured that even if at first they were not entirely at ease with each other, they are now indissoluble; an oil painting on canvas, on the other hand, is made, one could almost say, to be displaced, to be a *sans patrie*, and it does not much matter which great museum it ends up in or who are its neighbours.

And to it, in its new home, will come the tourist group, being lectured to by a guide. What is it that goes through the minds of its members as they stand there, in the Louvre, gazing at the *Mona Lisa*, or, two days later, in front of Rembrandt's *Night Watch* in the Rijksmuseum in Amsterdam? Presumably that they have seen a masterpiece and that this has somehow made them more cultured, more knowledgeable, perhaps even, who knows, better. But what is such culture or knowledge worth? What has it opened in their minds and hearts?

'Please do not touch.' That is the one label every visitor to the museum or gallery has read. And if one forgets its message, if one leans forward a little too closely, puts out a hand to follow a contour or point a figure out to a companion, there is always an attendant on hand to warn one away. But, in a sense, the label is both redundant and misleading. In a museum or gallery all the great and famous objects of world culture are 'at hand'. For the duration of our visit they belong to us. We can, in the British Museum, move from the Rosetta Stone to the Elgin Marbles, from the Lindisfarne Gospels to some ancient Chinese buddha. It is an admirable place to find visual confirmation of what books have told us, and in that sense it is an invaluable educational tool. But just as the gramophone and the radio have brought the masterpieces of world music to us without our having to make any effort to get to them, so here the very abundance and proximity of masterpieces and objects of huge cultural significance tend to deprive each of its aura. We can, if the attendant is not looking, actually touch them – but can they touch us?

3   Boundaries

The trouble with mirrors, as Merleau-Ponty notes in *La Prose du monde*, is that they show too much. I do not see my body in the ordinary course of things as I see it in the mirror. It is not an object laid open to my gaze, as it is in the mirror, but that which looks, feels, moves. The world exists for me not because I see it but because I am a part of it.

In the ordinary course of things I do not look, I merely take in. But once there is a frame around my field of vision my relation to the field changes. What is within that frame is immediately arresting, it asks to be examined. At the same time what is within the frame is cut off from the rest of the world and cut off from me. The peculiar horror and fascination of mirrors lies in the fact that they present the world to us not as we normally experience it but as both open to our gaze and yet forever beyond our reach.

In our daily dealings with the world we do not encounter frames and we do not gaze. As I talk to a friend it is not his face or body that holds my attention but simply himself. When I shake hands with someone, as Merleau-Ponty again notes, I am conscious not of grasping a hand, flesh and bone, but of meeting someone. It is the same with speech. I do not analyse my friend's words in order to try and understand what he is saying, I merely grasp his meaning. When I am reading a book I do not read words, I read the book; when I am looking at a painting I do not see brushstrokes, I see the painting. Of course the book may direct my attention to its words, the painting to its brushstrokes, just as my friend may make a pun or quote a poem, but that does not alter my essential relationship to the book, the painting or my friend.

At the same time it would be wrong to imagine that even my encounter with my friend is a totally natural occurrence. For it to work as it normally does we both have had, over the years, to learn the rules that underlie such events. For example, I will only respond to him rather than analysing his words and behaviour if he speaks to me in the language we normally use together and if he behaves in a predictable manner. Were he to start talking Italian, for instance, or to stand on his head while talking to me, I might still be able to understand him, but I would not be able to carry on a conversation with him. Instead, I would be trying to analyse his words and gestures in an effort to understand what had got hold of him. Only if he sits down and accepts the cup of coffee I offer him can we pass beyond words and gestures to a conversation.

In other words, even encounters with friends take place within the framework of what literary critics call genre. The genre gives us the ground rules and the horizons of expectation for the encounter and so allows it to flourish. The genre of the conversation over coffee is different from that of the sherry party or the long-distance phone-call. We are not aware of the genre precisely because it is the ground of the encounter, made up not of rules which have to be learned but of practices which we have unconsciously come to master over the years.

It is the same with literary genre. Homer's initial 'Sing, Goddess, of the wrath of Achilles the son of Peleus', allows the audience to filter out the other possible things the poem might be about to do and settle down to listen to *this*. He may of course go on to flout or subvert or extend the genre, but that can easily be grasped, since the audience knows what the conventions are in the first place, just as my friend may in his excitement leave his coffee undrunk and get up from his chair to walk about the room, which will alert me to his state but not leave me bewildered as to his motives or intentions. Moreover, in the original setting, the audience would have known even before the bard started to sing just what kind of an evening they were in for, just as those attending the dramatic festivals in which playwrights were expected to put on a tragic trilogy followed by a satyr play would have known that Aeschylus was not going to give them a comedy or an epic recitation. That the conventionally silent third character on stage, Pylades, is suddenly given speech by Aeschylus and made to utter a crucial phrase in the second part of the *Oresteia* trilogy ('Make all men living your enemies, not the gods'), a phrase which finally persuades Orestes that he must kill his mother, would have made that moment more powerful, not incomprehensible.

Genre, in earlier literature, established the context in which the work was to be received. It was what made you feel you were in the presence of a friend, not a stranger, of someone to whom you could respond as a person, not someone whose every word and gesture had to be monitored for fear of misunderstanding. But of course this only works so long as the conventions on which genre relies are generally accepted. Already in the ancient world there is evidence that genre conventions could be misunderstood and thus seen as absurd or misleading or restricting. Plato's overt attacks on Homer and the covert attacks of Euripides on Aeschylus and Sophocles reveal what happens when the traditions and conventions underlying genre are no longer felt to be natural. But the most striking example comes, significantly, not from the Greek world but from the world of the Hebrew Bible. In Ezekiel 33 we find the prophet berating his audience for listening to him as they would to a bard or entertainer and not as to one who is trying to tell them an awful truth. 'And lo,' he says, miming their response, 'thou art unto them as a very lovely song of one that hath a pleasant voice, and can play well on an instrument: for they hear thy words, but they do them not' (Ezekiel 33:32). But of course we are to imagine his audience clapping even harder at this, thus leaving the prophet helpless as to how to convince them that what he is saying is not art or entertainment but a true account of what will come to pass.

It is as though the conventions of art, instead of enabling discourse to take place, had come between Ezekiel and his audience. He wants to sweep them away but this turns out not to be possible. Once he starts to talk to his audience the possibility of misunderstanding will always be there and nothing he can say will get rid of the problem. The possibility had always been there (and is probably there in any culture), but it only became manifest in a time of cultural and epistemological confusion and change. We find it in Plato and in Euripides; in the Hebrew prophets and in St Paul; but it only comes to dominate all other issues at the time of the Renaissance and the Reformation and their aftermath.

Two episodes are emblematic.

The first is to be found in a book. When, in Rabelais' *Pantagruel*, the young giant meets Panurge, who is to become his lifelong friend, a curious scene ensues. Pantagruel asks the bedraggled but handsome youth who has appeared before him who he is and where he comes from. Panurge answers first in German and then, when this draws a blank, in Italian, Dutch, Spanish, Scots, Hebrew, Greek, Latin and a couple of other, unidentified, languages. Pantagruel is totally baffled. Finally he asks Panurge: 'Say, my friend, can you not speak French?' 'But of course, sire,' answers the other, 'it is my natural and mother tongue.' Then, says the giant, tell us your name and where you come from, for, he adds, I have taken such a liking to you that I would make you my companion. At which point Panurge, in perfect French, gives him all the information he has asked for, but suggests that the full story of his adventures had better wait till he has eaten and drunk, for his body is crying out for nutriment. Upon which Pantagruel gives orders that he be brought to his lodgings and royally entertained and thereafter they are inseparable.

The comedy here lies in the fact that Panurge acts and speaks as though he does not know the conventions of social intercourse. There is after all no law which says you have to answer someone in the language in which they address you, even if that language is French and you are in France and yourself happen to be French. It is only, as they say, that it helps.

But if Panurge is a joker he is, like Shakespeare's fools, also making a serious point. After all, anyone can beg – why should we believe him when he says he is hungry and thirsty and needs clothing? By couching his plea in every language but French Panurge not only gets the chance to reiterate his plea until it seems to take on the force less of an argument than of a physical fact, but he also brings out into the open the point that there will always be a gap between what we say and how we feel, that no language on earth can *express* our bodily needs. And the young king seems to understand, which is why in the end he offers him much more than food and clothing, he offers him friendship.

Panurge's name tells us that he can do all things, that he is a fixer; and his account of his life reveals that he is a vagabond and an adventurer. Where Pantagruel and his court are rooted in a specific place and specific conventions, Panurge is rootless, without history or genealogy, a crosser of boundaries, a man who lives by his wits and his wit. That is also why he fascinates Rabelais. Indeed, we could say that the sudden appearance of Panurge in chapter nine – and the first edition of the book prints two chapter nines, suggesting that Panurge's eruption was indeed sudden and unexpected – coincides with Rabelais' discovery of his vocation.

Consider. *Pantagruel* was Rabelais' first venture into an as yet unknown field, that of printed prose fiction. When he begins Rabelais has as yet no defined public, such as Dante and Petrarch had, nor does he have a patron to support him and instruct him as to what he requires. It is going to be up to him to decide what to write and to draw out a public from among the readers of both popular romances and highbrow epics. Thus, like Panurge, Rabelais is both free and needy. Before chapter nine he had been content to advance by means of parody of both epic and romance, making us laugh at his implicit message – this is not epic and it is not romance, the days for such things are past – but not really sure of what his own territory or voice should be. After Panurge's arrival everything changes.

I am not a bard, Rabelais is insisting. I am not the spokesman of the community. Words have, in a printed book, no authority whatsoever. But what role does an author have then? The European novel (novel: new, as *roman*: vernacular) develops by making the boundaries evaporate, by asserting, with varying degrees of seriousness and the desire to persuade: this is true because I, Robinson Crusoe, because I, David Copperfield, tell you it is true. But Rabelais takes a different path. He does not wish merely to entertain, like the authors of the old romances and their descendants, and he is too proud of his new-found freedom to wish to spend his time and powers pretending to be bound by the constraints of the memoir, even the invented memoir. On the other hand mere irony, the mere debunking of epic and romance, does not satisfy him any more than it will satisfy Cervantes in the following century. What Panurge helps him to discover, and what Cervantes, Sterne and Beckett will discover in turn, perhaps with a little help from him, is this: that through laughter a certain truth will emerge, behind the playful words a message will be heard: I am hungry; I am needy; I want to speak to you; I want you to listen to me; yet the words I have to use will always be someone else's words, the language will always be foreign to my own specific needs. Precisely because we can never pin the jester down, precisely because he always seems to know himself and us so well, precisely because he makes us laugh, we extend to him our trust. We do not believe he is telling us true things about *his* life (Panurge's, Rabelais'), but rather that he is telling us something far more important: he is telling us what life is like in a world where all boundaries have become fluid, and he is showing us the falsity and dangers of pretending that this is not the case.

The second episode is not fictional but historical, not comic but rather sad. When, at the end of the eighteenth century, Dr Johnson, in his bluff way, censured Milton for using unnatural and high-flown language to mourn the death of his friend Edward King, he made a point about *Lycidas*, but he also, unwittingly, made a point about himself and about the fatal divisions that were beginning to emerge between segments of the literate public. For him there was something deeply unsettling about Milton's style and chosen form, since the death of a friend, above all things, should have elicited direct and genuine emotion. But Milton, of course, was not concerned to be 'sincere'. For him the adequate expression of his feelings for King could only be realised in the form of a pastoral elegy.

Since the time of Dr Johnson critics and readers have been divided as to whether Milton was right and Johnson wrong, or whether Milton was, in effect, hopelessly out of date and trying to do something which could simply no longer be done, as Johnson sensed.

For, as we have seen, already a century before Milton Rabelais had come to accept – and to turn to his own account – that literary genre, being no longer universally accepted (that is, underpinned by a shared view of the world and shared traditions), needed to be redefined and could no longer go on being used uncritically. At roughly the same time as Milton was writing *Lycidas* Molière was offering his audiences a play about a man who has no use for the polite conventions of society, and preachers everywhere were urging their congregations to reject the traditions of reading that had grown up round the Bible and instead to respond directly to the naked word of God. Wherever we turn it is the same story. In the introduction to his *Principles of Human Knowledge* (1710), George Berkeley writes that he intends to avoid error by taking ideas 'bare naked' into his mind without even the mediation of words; and Defoe at almost exactly the same time was setting the novel on its course by presenting his fictions not as the artefacts of a maker or stories told by a story-teller, but as the unadorned and direct account of his sailor and prostitute protagonists.

But, as with film, the greater lifelikeness of the novel is bought at a cost. Here too it is now possible to experience vicariously the most extraordinary adventures, to live within the skin of a Robinson Crusoe or a Moll Flanders for the duration of the novel. But here too this is possible only because the body of the reader has, so to speak, been left behind. (Not quite as much as it is in the cinema, for the words still have to be formed, even in silent reading, thus to some extent engaging the reader actively, but certainly far more than in the reading of poetry or of the prose of Rabelais, Nashe and Sterne.) But this fact had to be repressed or the benefits would not follow. By the nineteenth century, however, the more aware practitioners of the new form, such as Poe, Hawthorne and Dostoevesky, began to sense that a price would have to be paid for this act of repression. Their works are haunted by mirrors and by that creature of the mirror, the double, as though they would allay the guilt of repression by turning its consequences into their actual subject-matter. Poe's 'William Wilson', which contains a double whose very voice echoes that of the protagonist, and which ends with a murder-suicide in front of a mirror, merely makes manifest in a rather melodramatic way the tensions hidden in the new form, and suggests that if there is to be an honest ending it will have to be self-annihilation.

What is troubling about mirrors, I suggested, is that they demand to be looked at. We cannot live with a mirror as we can with sofas and tables, merely accepting it as a part of our daily environment. But has not painting, since the Renaissance, forced us to look at it head-on, as though it *were* a mirror, to stand back and examine what is within the frame rather than simply living with the image, as the worshipper had done for centuries with altarpieces and stained-glass windows?

To compensate for this the greatest artists have taken a variety of steps. Chardin, for example, in his still lifes, shows us the worn and weathered surfaces of pots and pans and jugs, making us aware simultaneously of the beauty of the quiet craftsmanship that has gone into their making, and of the comparable labour that has gone into the painting we see before us. But the main device painters have used to compensate for the insistent gaze to which they know their work will be subject is to let a window into their painting in order to draw our gaze out into the distance. The window is a space within the painting which is free of the tyranny of gaze, a framed space within the larger frame of the painting

24

which acts as a focus for the release of tension generated by the larger frame and painted surface. It is fascinating to note that in two of the greatest paintings which herald the triumph of the accurate depiction of realistic space, van Eyck's Arnolfini double portrait in the National Gallery and Velázquez's *Las Meninas* at the Prado, the key elements are the mirror and the window or door. Foucault has written eloquently on the latter as a pivotal work in the transition from medieval to modern, but it is the former which is perhaps more interesting from our point of view.

At the back of the room, behind the rather inward-looking yet hieratic figures of Jan Arnolfini and his pregnant wife, a convex mirror hangs on

1. Jan van Eyck, *The Marriage of Giovanni (?) Arnolfini and Giovanna Cenami (?)*, 1434. (National Gallery, London)

the wall. In the space of the painting it nestles just above the barely touching hands of the two protagonists, too low in realistic terms to reflect what the painter in fact makes it reflect, yet at the precise point at which the perspective lines of the composition draw the eye. Had the wall at this point been blank it would have left us frustrated and trapped; instead, as we look into the mirror, we are drawn back once more into the room, and a perpetual movement is set up, for the room and its inhabitants in turn draw us back to the mirror. As we look into the mirror we see the backs of the two protagonists and, on the threshold of the room, two other figures and a gleam of light through the open door. The mirror also manages to reflect the window which we see on the left of the picture, with an orange-tree growing in the garden outside and four oranges, one on the window-sill and three on a table just beneath the window.

But the mirror itself is also an object in its own right. Its inner frame is circular and picked out with dabs of blue and red; outside that, in ten little panels, are depicted ten scenes from Christ's Passion, from the Agony in the Garden to the Resurrection, with the Crucifixion at the very top. Above the mirror, in a most elaborate script, the painter has inscribed the words: 'Johannes de Eyck fuit hic, 1434'. Art historians have often commented on the boldness of the gesture: a painter signing his own work – itself still unusual at that date – not in a corner but in the very centre of the picture. Some have suggested that it is the artist's witness to the marriage of the Arnolfinis, but while this may well be the case it still does not account sufficiently for the fantastic elaboration of the script. What has not so often been noted is the teasing ambiguity of the inscription: 'Jan van Eyck was here, 1434'. But where is 'here'? In the Arnolfini house? On the canvas? The phrase is like the mirror, in that it sends the reader back into the picture and then from the picture to itself. *Fuit* – was – asks us to ponder on the fact that Jan van Eyck was and is no more. That, along with Jan Arnolfini and his wife, he is long since dead. That the inscription, like the portraits, is only the trace of a life that once existed but does so no more.

At the same time the extraordinary richness of the initial 'J' activates in the viewer the very sensation of the calligrapher at work. Like the paradox of the Cretan liar it denies what it asserts and asserts what it denies. For, writing 'Jan van Eyck was here, 1434', and writing it so beautifully and so boldly right in the centre of the picture and just above the tiny scene of the Crucifixion, the painter ensures that he *is* in some sense still here, today, brought to life again by the viewer whenever the painting is looked at, just as for the believer Christ crucified rises again at every performance of the Mass.

In the same way the mirror beneath the inscription both affirms and negates the effect of mirrors, reproducing for ever an image that can only have been fleeting and asking us to note the act of artistic making without which it would not exist. Looking at the painting we experience wonder: wonder that such a work could have been made, wonder at the fact that Jan Arnolfini and his wife and Jan van Eyck existed then and left these traces, and wonder too that we exist and are seeing this, here, now.

4 Holding and Grasping

Every day the prisoner in solitary confinement is confronted with the same finite set of objects: bed, chair, table, walls, door, bucket. He has seen these so often, has run his hand over them so often, that he knows them by heart. Because he knows that today they will be the same as they were yesterday and that tomorrow they will be the same as they were today he finds that they have lost the power to look back at him. By contrast, the room in which productive work is being done, the room one enters with anticipation each morning and which one is free to leave at will, grows sanctified by use. Unnoticed, taken for granted, it is nevertheless felt to be beneficent, almost, for the one at work in it, a blessed space.

That is why for the prisoner, a window, however small, is so important. For the sky is never the same from minute to minute or from day to day, and there is always the possibility that a bird will fly across the sky as the prisoner looks out. Moreover, the sense that this is the same sky as stretches over those who are free gives the prisoner daily hope, reminds him every day that he is still alive.

That is why, too, daily exercise is so important to the prisoner and, if possible, books to read and pen and paper with which to write. Otherwise the lack of movement, the lack of conversation with others, the sheer monotony of his existence, may gradually lead him to feel that he has lost his own body, and the sense of that loss may make him lose his mind.

Of course when there is nothing to do there is always the wall to run your finger over. But the prisoner touches the wall of his cell only to remind himself of what keeps him shut in. Or perhaps he is driven by the faint hope that his fingers may tell him what his eyes cannot, that a crack has appeared in the wall, or that the crack which was already there has grown so large that he may just be able to squeeze through it to freedom.

He does not merely touch the wall: he examines it, he explores its surface, he tests its firmness.

But it is as he thought: nothing has changed. The following day, however, for want of something to do, he starts to examine it, to test it again.

If I touch what I have touched a thousand times and seen a thousand times it is as though I had not touched it. On the other hand if, with my eyes closed, I touch something that I cannot recognise, I will pull my hand back sharply in fear and revulsion.

But what of the familiar feeling of the garden wall under my hand as I leave the house at dawn to walk out into the hills? What of the familiar feel of the tennis racquet in my hand as I prepare to serve? Of the hockey stick or cricket bat as I take the field?

Clearly it is not just a question of familiarity. It has something to do with trusting things, with taking them for granted as allies and companions rather than as enemies and obstructions. If the garden wall were not there, reassuringly, under my hand, I would feel that something was wrong. If the racquet handle felt unfamiliar my service would disintegrate. Wall and racquet are here like the mother's hand which the child instinctively takes as he sets out for a walk. He does not take the hand to confirm anything or to test anything; he merely takes it because that is what one does at the start of a walk.

Yet if the hand is withdrawn his world collapses. Suddenly that hand, to which he had never given a thought, becomes the most important thing in the world to him. He realises he needs it the moment he has lost it. He knows now that he has to have it. He lays siege to his mother till she returns it to him. Yet once he has got it back his attitude to it has changed. He no longer merely holds it; knowing that it could at any moment be withdrawn, he now clings to it.

What had always been there when he wanted it has suddenly been withheld. It has been withheld for reasons the child cannot fathom. And it becomes important, crucial even, that this should not happen again.

29

The owner of that hand, instead of being the unseen, unthought-about ground of the child's universe, now becomes an opponent, and calculations have to be made as to the best means of defeating that opponent. Should the owner of the hand be threatened? Or cajoled? Or both?

It is no longer enough, though, merely to regain the hand a second or third time. What the child needs is for the *status quo ante* to be regained, for a situation to be put in place whereby the devastating arbitrary withdrawal of the hand will never again take place. To that end he bends every thought, to that end he strains every fibre of his being.

Yet the more he clings to the hand he has now got hold of again, the more restless will the owner of the hand become. The mother had been happy to hold her child's hand, but now that he has started to cling she becomes less happy. She feels that perhaps this excessive dependence is unhealthy, that it can only lead to sorrow in the future; or she suddenly resents the child's assertion that he has a right to her hand at all times. There is, probably, a mixture of both, for she is as confused by the emotions she has unwittingly aroused as is the child himself. The fact is that, suddenly, in the place of life being simply lived, there is drama.

No one has explored this abrupt and violent transition and its consequences better than Proust. One could almost say that *A la recherche*, for the whole of its enormous length, explores nothing but that.

His life as a unique individual begins for Marcel with an episode very similar to the one I have just described. He has gone to bed as usual and, as usual, is expecting his mother to come up and kiss him goodnight. But Swann has come to dinner and his mother is too busy to appear. Marcel grows more and more anxious; eventually he can stand it no longer and sends a servant with a message to his mother, begging her to come. She does not reply. Confusedly he determines to wait up for his parents and waylay them on their way to bed. He carries out his plan, confronting the surprised and sleepy pair on the stairs with a kind of heroic despair, for while he longs for his mother's presence he also dreads his father's anger. But there is no other way: he cannot, at this stage, hope to catch his mother alone.

To his surprise it is his father who gives in at once to his entreaty, while his mother, perhaps out of fear of appearing weak and indulgent in front of her husband, tries to refuse. But the father is adamant: 'Go on. Stay with the boy. Can't you see he's overwrought?'

So Marcel has won. His mother settles down in his bedroom, he climbs back into bed, and she begins to read to him from one of his favourite books, George Sand's *François le champi*. Her familiar voice soothes him and his anguish begins to abate. But now he discovers, to his sorrow, that in the new world he has entered every victory is also a defeat.

But how can this be? Marcel has triumphed. He has made his plans and carried them out as boldly as any Caesar or Napoleon. As a result he has acquired his objective: his mother is there with him in his room. But the fact that he has had to scheme for this has ineradicably affected the situation. He looks at his mother and sees her as if for the first time. She is no longer the ground of his being but a person in her own right, someone with a will and desires of her own, yet capable of being influenced by external events. And what he sees is a woman no longer young, a woman with a few tell-tale grey hairs already on her head – in other words a woman subject to time, a woman being carried inexorably towards the moment when, whatever he may do to try and stop it, she will no longer ever come to him when he calls.

His victory over her has made him understand for the first time that the only victor in such conflicts is Time, and that once we engage in combat with him we can only ever lose. Such understanding, however, is worse than useless. This scene is to be repeated with almost monotonous regularity in the course of Marcel's life. What he desires from every woman he falls in love with is that she should come to him of her own accord. But since this does not happen he has to use all his guile and seductive skill to win her. Having done so, though, he finds that it is no longer the same woman. His prisoner now, she is no longer the person he so desperately desired. And the more he tries to cling to her, the more he tries to bind her to him, the more anxious she becomes to escape. And the more he senses her anxiety, the more he clings to her and tries to hold her. Inevitably he loses her, as he had, that fateful night in Combray, in effect lost his mother.

The world for Marcel before that night was a world of happy iteration. Days went by, each new and exciting precisely because it did not threaten to be in any essential way different from all the others. Because it was a world grounded in trust it was a world suffused with aura, a world of reciprocity, in which what you looked at looked back at you. After that night it is as if Marcel had been projected violently into time and change, desire and frustration. What had simply been life had suddenly become a drama, a story, *his* story. That is why the details of that fateful night constitute his first clear memory, and that is why this is where the novel proper starts. Not just this novel, of course, but the novel as a form, a way of making sense of experience. The novel, this novel suggests, is always in search of a lost paradise, of the paradisal state which existed before it was propelled into existence, and it will not rest till that state is found again.

But *this* novel is a little different. After all, it does not start with the mother's withholding of her goodnight kiss, but earlier and elsewhere. And when iteration gives way to linearity Marcel finds that he has fallen (so to speak) into his life, that unique life which can only be his and no one else's; but also that, in a curious secular echo of the myth of the Fortunate Fall, his sense of loss is precisely what drives him and enables him to write; and writing, he will discover, many thousands of pages later, is what will allow him to overcome, at least partially, the trauma of that night, to draw the scattered pieces of himself into a whole, even as he recognises that such wholeness is only a dream of wholeness, something he can reach out for and touch but never actually grasp.

5   The Room

When, in a room by ourselves, we hold one hand in the other, we do not call that 'holding hands'. When, in a room by ourselves, we reach out towards our hand in a mirror and meet only the coldness of the glass, we do not call that touching. On the contrary, both are a sour parody of touch, born of and fuelling our sense of dejection, the sense of existing in a world which remains stonily indifferent to our needs and desires.

Yet there is, of course, one kind of solitary touching which does seem to bring one back into the world or bring the world back to one, and it is hardly surprising that it is Proust, once again, who has explored this with the greatest acumen and tact. It is just after Marcel has discovered to his sorrow that 'identical emotions do not spring up simultaneously in the hearts of all men in accordance with a pre-established order', that he recounts how '[s]ometimes to the exhilaration which I derived from being alone would be added an alternative feeling which I was unable to distinguish clearly from it, a feeling stimulated by the desire to see appear before my eyes a peasant-girl whom I might clasp in my arms'. Pursuing this memory he comes to understand that the woman he thus desires to hold in his arms, if she was in some sense brought into being by the woods and meadows through which he wandered daily, was also, in a sense, the incarnation of those woods and meadows, a single living being through whom he might possess the entire landscape. 'For at that time', he says, 'everything that was not myself, the earth and the creatures upon it, seemed to me more precious, more important, endowed with a more real existence than they appear to full-grown men.' In desiring a woman one does not then, as in later life, think of the pleasure she will give us, 'for one does not think of oneself, but only of escaping from oneself [car on ne pense pas à soi, on ne pense qu'à sortir de soi]'.

Soi, oneself, is the self when it feels dead, cut off from the source of life, even though for the young Marcel this is more an itch of the body than a conscious attitude, as it was for Wordsworth and Coleridge in their moods of dejection. It is the condition of knowing that one is alive but not being able to feel it, of feeling rather that there is life but it is elsewhere and that one is somehow cut off from it. That vague unfocused longing, which feels as if it would be appeased by the touch of another is so frustrating precisely because it seems as though so little is required to bring it to fulfilment, yet that little is nothing less than everything. Marcel may long for the peasant-girl with every fibre of his being, but she does not appear. 'Alas,' he says,

> it was in vain that I implored the castle-keep of Roussainville, that I begged it to send out to meet me some daughter of its village, appealing to it as to the sole confidant of my earliest desires when, at the top of our house in Combray, in the little room that smelt of orris-root, I could see nothing but its tower framed in the half-opened window as, with the heroic misgivings of a traveller setting out on a voyage of exploration or of a desperate wretch hesitating on the verge of self-destruction, faint with emotion, I explored, across the bounds of my own experience, an untrodden path which, for all I knew was deadly – until the moment when a natural train like that left by a snail smeared the leaves of the flowering currant that drooped around me.

Rarely have the dynamics of desire and addiction been better caught, not even in the classic accounts of alcoholism and drug addiction, such as *Under the Volcano* and William Burroughs' *Junky*. In the mirror I contemplate myself as nothing but surface, yet what I need if I am to come alive is for a path to open up that will lead me into myself and so back into the world. That is what drink seems to be able to do with the Consul in Lowry's novel, and that is what masturbation here seems to do for Marcel. But the conclusion of the episode provides us with the inevitable corollary: what seems to be a means of bringing us back into touch with the world, a particularly satisfying means, since we can control it ourselves and are not therefore dependent on the whims of others, turns out to alienate us ever more from the world as the means become an end in itself. 'I ceased to think of those desires which came to me on my walks, but were never realised, as being shared by others, or as having any existence outside myself', writes Proust. 'They seemed to me now no more than the purely subjective, impotent, illusory creations of my temperament. They no longer had any connection with nature, with the world of real things, which from then onwards lost all its charm and significance.'

It is one of the few novels of the second half of our century which can vie with *A la recherche* in ambition and execution, *La Vie mode d'emploi* by Georges Perec, that provides us with the final gloomy metamorphosis of the solitary man in the room. In one of the shortest sections of the book a room and its contents are described in just two paragraphs, the first of which reads: 'Today the room is occupied by a man of about thirty; he is on his bed, stark naked, prone, amidst five inflatable dolls, lying full length on top of one of them and cuddling two others in his arms, apparently experiencing an unparalleled orgasm on these precarious simulacra [semblant éprouver sur ces simulacres instables un orgasme hors pair].'

6 Addiction

Like a great many people I have been trying to give up smoking. In the process I have learned many things about myself and a little bit about the nature of addiction, but I have still not given up.

I know I am addicted because, whenever I try to imagine a life without cigarettes, without *any* cigarettes, *ever*, I realise it is a life I do not want.

In the early hours of the morning, when I cannot sleep and my throat is dry and my tongue feels too big for my mouth, I promise myself that I will pull myself together and stop the very next day. After all, I have quite a strong will, would, when young, push myself to get up at six in the mornings to get a bit of extra swimming or running training in when a big race was coming up, and later forced myself to work for exams in subjects in which I was not interested. Surely giving up something must be easier than forcing oneself to do something? But when the next day comes I do not stop. And the cigarettes I smoke in the course of the day are not an indulgence, like eating chocolates, say. They are a necessary part of my life. Without them – and I am not a great smoker, rarely consume more than eight or ten cigarettes a day – without them I simply do not feel alive.

It's as simple as that. Were I to give up I would of course be able to go on carrying out the tasks I have to do.

But I feel I would do them like an automaton. And the deprivation I experience when the need for a cigarette comes upon me is not like the sense of the simple absence from my diet of food I like or even need. It is more like being in solitary confinement. To smoke a cigarette has much in common, for me at any rate, with going for a walk: when I put a cigarette in my mouth and light up, my body, which, until then had grown dead, unresponsive, begins to come to life again and I feel myself to be once more a part of the world.

They say that the famous advert, 'You're never alone with a Strand', killed that particular brand of cigarette stone dead. One must never, it seems, associate a product with solitude. That may be advertising wisdom, but I have never come across a truer description of what smoking means. To take out a cigarette, put it in one's mouth, light it, inhale, exhale, is to feel the deprivation of solitude dropping away. To smoke a cigarette is like sitting by an open fire, which is as much as to say that to smoke an ersatz cigarette, one of those nicotine-free monstrosities designed to wean you from the habit, is like trying to sit down to relax in front of a gas or electric fire, perhaps one of those unspeakable contraptions that pretend to be a coal or wood fire.

Since I cannot imagine a life without cigarettes it is clear that, however much I may try to give up, day by day, I will not be able to do so. Only if I could actually *want* a life without cigarettes would I be able to give up. But what I want is a life without the unpleasant consequences of smoking, and that is quite a different matter. Auden has some interesting things to say on the whole subject in his notes to *New Year Letter*. Glossing the line 'Hell is the being of the lie', he says:

> It is possible that the gates of Hell are always standing wide open. The lost are perfectly free to leave whenever they like, but to do so would mean admitting that the gates were open, that is, that there was another life outside. This they cannot admit, not because they have any pleasure in their present existence, but because the life outside would be different, and, if they admitted its existence, they would have to lead it. They know this. They know that they are free to leave and they know why they do not. This knowledge is the flame of Hell.

This suggests that if we want to find the fullest anatomy of addiction yet compiled we need to turn not to a medical textbook but to Dante's

Inferno. Here the damned cling to their past lives as in those lives they clung to themselves. In the upper reaches of the cone of Hell they go endlessly round and round, buffeted by winds or with fire raining down upon them; in the lower circles they grow more and more immobile until, almost at the centre, they are frozen into the ice so that even their tears can no longer flow. In Hell the basic premise is adumbrated by Virgil early on: 'sanza speme vivemo in disio' – 'without hope we live in longing'. In Purgatory, on the other hand, though the going may, especially on the lower slopes, be exceedingly hard, the pilgrims live always in the hope that one day they will reach the top of the mountain and thence be released into Paradise. The adjective 'green', descriptive of the new shoot, and the verb 'to turn', implying spiritual as well as mental suppleness, are the key words in the canticle of Purgatory, for those who were supple enough in spirit to turn to God in repentance before their death, to admit, in Auden's words, that there is another life outside, now find themselves able to move from one level of the mountain to the next, shedding as they do so the burden of their guilt, returning to their true selves.

Tornare is linked in Dante's poem not only to *verde* but also to *amore*: love is what makes the world go round, quite literally, as well as what makes the movement of Dante's verse possible. Those in Hell who refuse love are unable to turn (the Hebrew word for repentance is *teshuvah* – turning) in their lives, and now in eternity they remain locked in their various circles and in the bed of torment that is their own bodies.

Canto five of the *Inferno*, so often taken as a classic account of the nature of romantic love, whether condemned or admired by Dante, can also be seen as a profound exploration of the nature of addiction. Its abiding power stems from the fact that we cannot simply pity and condescend to the two lovers, Paolo and Francesca, but have to recognise that they present us with a pure and extreme version of what we have probably all experienced at one time or another: the power over us of desire, the craving never to leave the side of the beloved, even the painful sweetness of letting the rest of the world go hang so long as one can cling to the being one loves.

Since one characteristic of addicts is their ability and perhaps even need to talk about and justify themselves, it is natural that those in Hell make the longest speeches. Francesca's, in this canto, is one of the longest. Dante and Virgil have entered a place 'mute of all light', where 'the hellish hurricane, never resting, sweeps along the spirits with its rapine; whirling and smiting, it torments them . . . hither, thither, downward, upward, it drives them. No hope of less pain, not to say of rest, ever comforts them.' Dante notices 'two that go together and seem to be so light upon the wind', and asks Virgil if he may pause and speak to them. As they come past he addresses them graciously and asks them to converse with him. At once Francesca complies, recounting their story with a mixture of poetic elaboration, dignity and self-pity that is hard to resist:

> Siede la terra dove nata fui
>> su la marina dove'l Po discende
>> per aver pace co'seguaci sui.
> Amor, ch'al cor gentil ratto s'apprende,
>> prese costui de la belle persona
>> che mi fu tolta; e'l modo ancor m'offende.
> Amor, ch'a nulla amato amar perdona,
>> mi prese del costui piacer sì forte,
>> che, come vedi, ancor non m'abbandona.
> Amor condusse noi ad una morte.
>> Caina attende chi a vita ci spense.
>
> (97–107)

She was born, she says, where the river Po descends 'to be at peace with its followers'. That natural peace is clearly something she warms to and longs for, and in the next few lines, each *terzina* introduced by the powerful word *amor*, she almost manages to leave the impression that she has attained it. Love, here, is almost a god, Amor, who has taken her over. (La Pia, in canto three of *Paradiso*, will speak of having surrendered herself to the Christian God, 'who draws our wills to what He wills; and in His will is our peace' (84–5) – a passage Dante obviously means us to compare and contrast with this one.) This god, she tells Dante, 'absolves no loved one from loving', and it is he who 'brought us to one death'. The oneness is clearly the fulfilment of a profound desire, but how are we to understand the sentence as a whole? Is she blaming or praising the god for what he has done to them? Does she know herself? Is she not in effect allowing an extreme indulgence of the will to pass itself off as a necessity, as a total absence of will or choice? After all, her beautiful and melancholy words make no mention of the actions and decisions which must have been taken by the two lovers, and pass over the fact that she was married to Paolo's brother who, finding them together, killed them both. It is he who has been hurled into the circle of Cain for his deed, but she is too fastidious to spell this out or to admit that it is their own adultery that has brought them here, where Dante finds them – *come vedi* (as you see).

Yet, despite this duplicity or sentimentality (the two go together, since it is the belief that life has been hard on us that leads to self-pity), her description of the course her life has taken is profoundly moving, not least because of the conflict it makes manifest, and which she herself does not seem to be aware of, or cannot admit to herself, between the longing for the peace known by the river and the need to be with her lover and to keep affirming their innocence. Yet this conflict does come into the open when, pressed by Dante to say more, she admits (echoing Virgil and Boethius): 'Nessun maggior dolore/ che ricordarsi del tempo felice/ ne la miseria' [There is no greater sorrow than to recall, in wretchedness, the happy time]. Locked in the prison of her desires, she can only move helplessly between a wretched present and a past which grows more idyllic each time she recalls and retells it, but whose relation to the present she persistently refuses to acknowledge:

> Ma s'a conoscer la prima radice
>   del nostro amor tu hai cotanto affetto,
>   dirò come colui che piange e dice.
> Noi leggiavamo un giorno per diletto
>   di Lancialotto come amor lo strinse;
>   soli eravamo e sanza alcun sospetto.
> Per più fiate li occhi ci sospinse
>   quella lettura, e scolorocci il viso;
>   ma solo un punto fu quel che ci vinse.
> Quando leggemmo il disïato riso
>   esser basciato da cotanto amante,
>   questi, che mai da me non fia diviso,
> La bocca mi basciò tutto tremante.
>   Galeotto fu'l libro e chi lo scrisse:
>   quel giorno più non vi leggemmo avante.
>   (124–38)

Paolo does not kiss her, he kisses her mouth, just as in the story of Lancelot and Guinevere which they are reading it is not the queen who is kissed but the 'desired smile'. In addiction, as in the mirror, the self is fragmented and bits of the body float free of any unifying responsible self. That is why fetishism and addiction go together. And even here it is the book and its author who are blamed, while Paolo and Francesca remain the helpless victims of circumstances and their passion.

Yet it would be wrong to think we could stand back from Francesca and simply 'place' her in some objective scheme of things, for Dante is no more dismissive of her than the narrator in Proust's novel is of his young self. The pity Dante feels for her, culminating in his swooning as she finishes telling her story, may in part be seen as a weakness and worthy of the reprimand Virgil delivers, an importing of purely human values into the immutable world of the afterlife, but Dante's poem does not work by denying human feelings but rather by incorporating them into some larger design. Belaqua Shua, in Beckett's early story 'Dante and the Lobster' is quite right to question Virgil's chilling phrase, 'Qui vive la pietà quand'è ben morta', which might be translated as 'Here piety lives when pity is quite dead' (xx.28), but at the cost of separating out too neatly the two senses of *pietà* in Dante's Italian. After all, in the *Convivio* Dante had described *pietà* as the greatest of the virtues and not simply an unthinking emotional response: 'Pity, however,' he says, 'is not an emotion, but a noble disposition of spirit, ready to receive love, misericord, and other beneficent feelings' (II.x.). For the truth of the matter is that Francesca is a mirror in which Dante can recognise much of what he most values in himself.

After all, Francesca, as critics have pointed out, speaks the language of the *dolce stil nuovo*, the language which Dante learned from his older contemporaries, Guido Guinizelli and Guido Cavalcanti and which, in his first great work, the *Vita nuova*, he made his own. And the important thing to note is that, in writing his mature masterpiece, the *Commedia*, he does not leave such language behind. After all, as he says to Bonagiunta in Purgatory: 'I'mi son un che, quando Amor mi spira, noto, e a quel modo/ ch'e ditta dentro vo significando' [I am one who when love breathes in me, takes note, and, in that manner in which he dictates within, go on to set it forth] (xxiv.52–4). This is the language in which he discovered himself as a poet and it is still the language in which, as Jill Mann has acutely noted, 'Dante realises his relationship with Beatrice in

Purgatorio and *Paradiso*'. 'It is of course true', she goes on, 'that he fills this language with a wealth of spiritual meanings. . . . But . . ., whatever the deepening of meaning, a continuity between earthly and divine love is maintained at the level of language; it is the language of the love-lyric that becomes the mould into which the spiritual experience is shaped.'

As with Proust in the little room smelling of orris-root, so here, what is presented to us is far too important, far too central to the narrator's own growth as man and artist, to be simply put to one side. The power of the work of both Proust and Dante stems from the fact that both recognise the immense significance of such experiences, incorporate them into their final works, yet manage, through and in their writing, to escape from their addictive hold back into the world of human choice and responsibility.

To say that addiction is a way of relieving loneliness is to give the wrong impression. What it relieves is the sensory deprivation attendant on solitary confinement. And solitary confinement, as I have been suggesting, does not require four walls, a locked door and a jailor; it merely requires that we lose our sense of natural reciprocity with the world and are so painfully aware of that loss that we try to make it good in any way we can. When William Burroughs questions himself as to why he ever became a heroin addict he has to admit that boredom had a lot to do with it. None of the options of bourgeois life open to him held out any interest for him and what he saw of those who 'got on', who made what the world considered a success of their lives, positively nauseated him. Becoming a junkie was a way of escaping from this, and he brings out powerfully how much of an effort of will is actually involved in that, how long it takes to get really and truly hooked.

Unfortunately addiction cannot fully satisfy our desires either and thus leaves us in the state which all those in Dante's Hell experience: the state of being in perpetual hopeless longing – 'sanza speme vivemo in disio'.

Stanley Cavell, in his thoughtful book on film to which I have already referred, comes close to making the same point about film but shies away from the identification of film and addiction. But his perceptive remarks cut across the banalities of most film theory and help us to understand what is at issue. 'To say that we wish to view the world itself', he argues,

is to say that we are wishing for the condition of viewing as such. That is our way of establishing our connection with the world: through viewing, or having views of it. Our condition has become one in which our natural mode of perception is to view, feeling unseen. We do not so much look at the world as look *out at* it, from behind the self. It is our fantasies, now all but completely thwarted and out of hand, which are unseen and must be kept unseen. As if we could no longer hope that anyone might share them – at just the moment that they are pouring into the streets, less private than ever. So we are less than ever in a position to marry them to the world. Viewing a movie makes this condition automatic, takes the responsibility for it out of our hands. Hence movies seem more natural than reality. Not because they are escapes into fantasy, but because they are reliefs from private fantasy and its responsibilities; from the fact that the world is *already* drawn by fantasy. And not because they are dreams, but because they permit the self to be wakened, so that we may stop withdrawing our longings further into ourselves.

Despite a number of obscurities, which the rest of Cavell's book do not wholly clear up, the central assertion here seems both profound and important. Movies seem more natural than reality 'not because they are escapes into fantasy, but because they are reliefs from private fantasy and its responsibilities'. What all the people in Dante's Hell have chosen, what Marcel so longs for when once he discovers that his mother may at any time withdraw and leave him, is the relief from private fantasy and its responsibilities. And movies, like sexual passion and masturbation, seem to free us, while we experience them, from our daily confusions, self-doubts and fantasies; they seem to put us in touch with a reality which is fully natural and which asks simply that we give ourselves up to it.

This way of looking at film also helps to make clear something Cavell touches on frequently in the course of his book but does not, it seems to me, ever quite bring into focus: the fact that for him (as for me, though I am a little younger than he is) the great age of movies was also the age of adolescence. (That our adolescence coincided with the great unselfconscious era of the Hollywood movie is of course pure chance, but it does suggest that we and all those born between 1920 and 1945 will always have a different sense of film from those born before or after.) Cavell, at any rate, spends a good deal of time wondering why the films he recalls with such pleasure are the middle-brow films produced with such confidence and in such quantities by the Hollywood machine, and why, now that movies seem to have caught up with the other arts, so to speak, and are dividing more and more clearly into high and middle-brow, there may be great films produced but it is impossible to view them with precisely the kind of delight with which we viewed them as teenagers.

Certainly my own memories of the excitement of movies, the excitement of the whole business of meeting up with my friends, going into town, buying my ticket, entering the vast palace, waiting for the music to stop and the lights to dim, then plunging totally into the world of the film, to emerge, three hours later, into the light of day, still lost in what I had seen – all this seems to have more to do with a particular moment in my life than with the quality of the films themselves. I am reminded of Graham Greene's comments on his childhood reading:

> Of course I should be interested to hear that a new novel by Mr. E. M. Forster was going to appear this spring, but I could never compare that mild expectation of civilized pleasure with the missed heartbeat, the appalled glee I felt when I found on a library shelf a novel by Rider Haggard, Percy Westerman, Captain Brereton or Stanley Weyman which I had not read before.

I suspect that childhood and adolescence, being a time of waiting and expectation, is also the time when the reading of such books and the viewing of films is natural and totally, uncritically pleasurable. Normally we grow out of Rider Haggard and of the missed heartbeat and appalled glee with which we used to watch whatever film we were allowed to watch. When we do not we can say that we have become addicted. Then we go to the cinema by ourselves and if the adventures of Robin Hood or Captain Hornblower fail to do the trick we may in time graduate to more and more violent and pornographic fare.

Addiction, then, in the full sense of the word, is an adult condition, though its seeds are sown in childhood. And so it may well be that it is far too solemn, far too censorious, to describe film as having all the ingredients of addiction, as I did earlier; it may simply be that film is a form that belongs, with the novels of Rider Haggard (and with *François le champi*), to a moment in our life when we are struggling to make the transition from child to adult, and that high-flown theories of the nature of film and the greatness of individual examples quite miss the point.

The people in ante-Hell blow neither hot nor cold and so are spewed out of both Hell and Heaven. Those in Hell are those who feel their condition as a lack, who desperately need to be relieved of private fantasy, its anxieties and its responsibilities, but who have lost or have never acquired a purposive sense of how this may be done in a way that will not leave them more miserable than before. Paolo and Francesca fall into each other's arms while reading the story of Lancelot and Guinevere because the passion of those story-book lovers seems to transcend and obliterate the destruction it leaves in its wake. That option too is open to us, even when we have left adolescence behind, though we more often prefer the less dangerous ones of smoking or drinking, or slinking off to the cinema by ourselves. It's no cure, of course, but, while it lasts, we are at least free of the burden of ourselves.

7 Transgression

> And it was then that something utterly unheard of,
> though, on the other hand something that in a certain
> sense was only to be expected, took place. The old man
> suddenly felt that instead of whispering some interesting
> secret to him, Nicholas all of a sudden seized the upper
> part of his ear between his teeth and bit it hard. He
> trembled all over and his breath failed him.

This is only the latest little joke Nicholas Stavrogin
indulges in in the little town where he has grown up. But
no one is in any doubt that more than a practical joke is at
issue here. The whole of Dostoevsky's *The Devils* is an
exploration of what becomes of boundaries when long-
held beliefs about religion and civilised behaviour start to
erode. The point Dostoevsky brings out so strikingly, both
here and throughout the rest of his mature work, though,
is how such actions, which advance no one and can only
rebound on the doer, are an endemic part of modern life,
where people feel they can only discover who they are by
crossing the boundaries of decorum and even of the most
profound moral taboos. Dostoevsky brings out in a
remarkably clear way both the horror and the pathos of
such transgression: horror because if anyone can do
anything, from biting the ear inclined towards him to
violating a child, then how can society function any longer,

how can we respectable citizens go about our daily tasks in peace of mind? Pathos because such acts are born out of despair, out of a sense that the world is no longer returning my gaze and I must provoke it to do so in any way I can or my life is simply not worth living. But such provocation will always fail, since it is instigated by me; indeed, crimes committed in this spirit require for their fulfilment that they be punished, for then at least the criminal will actually feel at last that the world is paying him some attention – a point made by the examining magistrate Porfiry to Raskolnikov half-way through *Crime and Punishment*.

These are not only heady and topical themes – today's papers are as full of crimes committed out of the desperate need to transgress as they were in Dostoevsky's day – they are also in danger of themselves turning into clichés, as law enforcers, psychiatrists and religious leaders all come up with ritualised answers whenever a particularly horrifying crime of this kind occurs. If Dostoevsky charted out the ground more than a hundred years ago it is Proust, once again, who can save us from succumbing to cliché, as he brings his steady gaze to bear on a scene of transgression even more banal than that of Stavrogin's biting of the governor's ear.

It is not by chance that Proust follows the episode of the 'little room that smelt of orris-root' with Marcel's first encounter with perverse sexuality: voyeurism, sadism, masochism and homosexuality all rolled into one. But as with the earlier scene what Proust shows is that perversity is merely one of the directions taken by the flowering of desire, which is itself understood as a consequence of the traumatic discovery that my needs and those of the world are not the same.

Proust sets up the scene as though it were a film, for the watcher sits in the shadows, unseen and safe, while the dramatic events pass in a lighted square before his eyes.

Marcel has strolled to the precincts of the house owned by the composer Vinteuil, who has recently died, and has lain down on a grassy bank. It is a hot day and he has gone to sleep. He wakes up to find that it is almost night. At an open window, practically on a level with him and barely a few feet away, he sees the composer's daughter. She has obviously just come in. She is in mourning for her father. On the mantelpiece is a portrait of him and, as Marcel watches, she goes over to fetch it. At this moment a carriage is heard coming up the drive and soon Mlle

Vinteuil's friend joins her. She moves to close the shutters. Her friend stops her. But someone may see us, she says. Here in the middle of the country? laughs the friend, and then adds: What if they do? All the better that they should see us.

So the window is left open and Marcel goes on watching. The two women chase each other round the room, then collapse laughing on the sofa. They embrace. Mlle Vinteuil says something about wanting to remove the portrait of her father from their vicinity, but her friend abruptly stops her: 'Let him stay there. He can't bother us any longer. D'you think he'd start whining, and wanting to put your overcoat on for you, if he saw you now with the window open, the ugly old monkey?'

They embrace again and then the friend takes the portrait in her hands and looks at it. 'Do you know what I should like to do to this old horror?' she asks, and whispers something to Mlle Vinteuil. 'Oh! You wouldn't dare.' 'Not dare to spit on it? On *that*?' asks the friend, with, Proust says, 'studied brutality'. Marcel, however, sees no more, for Mlle Vinteuil, 'with an air that was at once languid, awkward, bustling, honest and sad', does then get up and close the shutters. But, he adds, he now knows that, for all the sufferings his daughter had brought her father during his lifetime, there would be worse to follow for Vinteuil after his death. And yet, he says, with one of those typically Proustian additions which make one realise that genius is in effect nothing more than the ability to push an insight to its limits and not be deflected either by laziness or by conventional wisdom,

I have since reflected that if M. Vinteuil had been able to be present at this scene, he might still, in spite of everything, have continued to believe in his daughter's goodness of heart, and perhaps in doing so he would not have been altogether wrong. It was true that in Mlle Vinteuil's habits the appearance of evil was so absolute that it would have been hard to find it exhibited to such a degree of perfection outside a convinced sadist; it is behind the footlights of a Paris theatre and not under the homely lamp of an actual country house that one expects to see a girl encouraging a friend to spit upon the portrait of a father who has lived and died for her alone; and when we find in real life a desire for melodramatic effect, it is generally sadism that is responsible for it.

And yet, he goes on,

It is possible that, without being in the least inclined towards sadism, a daughter might be guilty of equally cruel offences as those of Mlle Vinteuil against the memory and the wishes of her dead father, but she would not give them deliberate expression in an act so crude in its symbolism, so lacking in subtlety; the criminal element in her behaviour would be less evident to other people, and even to herself, since she would not admit to herself that she was doing wrong.

And now, in a passage of great subtlety and profundity, Proust concludes:

Sadists of Mlle Vinteuil's sort are creatures so purely sentimental, so naturally virtuous, that even sensual pleasure appears to them as something bad, the prerogative of the wicked. And when they allow themselves for a moment to enjoy it they endeavour to impersonate, to identify with, the wicked, and to make their partners do likewise, in order to gain a momentary illusion of having escaped beyond the control of their own gentle and scrupulous natures into the inhuman world of pleasure. And I could understand how she must have longed for such an escape when I saw how impossible it was for her to effect it.

'Perhaps', ends Proust,

she would not have thought of evil as a state so rare, so abnormal, so exotic, one in which it was so refreshing to sojourn, had she been able to discern in herself, as in everyone else, that indifference to the sufferings one causes which, whatever other names one gives it, is the most terrible and lasting form of cruelty.

Indifference, not sadism, is the cardinal sin. Indifference implies a lack of imagination, an inability to feel what it is like to be another, and the result is cruelty, epitomised in the novel by the Duchesse de Guermantes' refusal to listen to Swann telling her he is mortally ill because to respond to that would mean the ruin of her evening out. We have all, at one time or another, acted like the duchess; what is hard for us to accept is that such actions place one in Hell no less than the deliberate treatment of others as vermin. For Hell has many circles and there is no essential distinction between the adoring Francesca clinging to her Paolo and the cannibal Ugolino gnawing at the skull of his enemy, between the duchess's refusal to hear what Swann is telling her and Gregor Samsa's family not wanting to accept that it is their son who has turned into a revolting insect.

Reading Proust, walking round museums and art galleries, listening to Beethoven, will not necessarily save us from such a fate. It may indeed be that such activities only reinforce our indifference ('In the room the women come and go,/Talking of Michelangelo'). Sadism, on the other hand, at least in the form practised by Mlle Vinteuil, like the masturbation explored a few pages earlier, represents a striving to escape the solitary confinement of indifference, a wild strategy to force an apparently indifferent world to touch us, if only for a moment. Unfortunately it can never truly succeed, for it is always we who instigate it and what we need is precisely the opposite: it is for the world to touch us, unawares. Thus it has to be repeated, over and over again, in ever more desperate and ineffectual efforts. But then how are we to act once we have fallen into the solitude of our lives, once the aura which had lit up the world and which we had, until one fateful moment, taken for granted, seems to have been lost for ever? Are we to wait, like Chaplin's flower-girl, for that miracle which may never happen? Or is there some other way in which we can help it into being, a less melancholy, less desperate way?

8 The Lesson of the Hand (2)

Colonus is said to have been Sophocles' birthplace, so that in *Oedipus at Colonus*, written when he was past eighty, the playwright is, among other things, celebrating the place where he was born. But at the same time he is celebrating the place which had been the central focus of his life, the Dionysian stage. Like Chaplin in *City Lights* he chose a blind protagonist, but he did so not to provide laughter or to jerk the tears of his audience, but to lead them to an exploration of the relations between those who watch a spectacle and the hero of that spectacle, between clinging and letting go, between the human body and the space it inhabits.

The aged Oedipus, who had blinded himself on learning that he had killed his father and begotten children on his own mother, has been cast out of Thebes and has arrived here, where we first see him, led by his devoted daughters, Antigone and Ismene. His first words put the theme of place squarely before us:

> Tell me, Antigone – where have you come to now
> With your blind old father? What is this place, my child?

She answers him not with a name but with a description:

> Here, where we are,
> There is a kind of sacred precinct . . .
> There is a seat of natural rock. Sit down and rest.

We must bear in mind that the Athenian stage is a clearly defined circular space, bare of props, fully visible to every member of the audience. As they wait for the play to begin, seated in the rising circular tiers, it is only a potential space, to be transformed into whatever the playwright wishes simply by virtue of his words playing on their imaginations. This play thus begins traditionally enough by setting the scene. What is startling is what Oedipus' blindness is already doing to the tradition. It is he, not just the audience, who is being informed about the locus of the play, and we are going to watch him, as he moves slowly round it, trying to relate what he feels with his outstretched hand to what he is told. In this way the audience, too, is made to experience the circular space it knows so well in quite a new way.

A man appears, a local countryman. He is horrified by what he sees: 'Sir, before you ask me any question, come from that seat. That place is holy ground.' Oedipus insists that this is where he was meant to come, and asks again: 'What is this place?' The countryman explains that it is Colonus, and, after repeating that it is sacred, adds:

> It is not such a place as is famed in song and story,
> But its name is great in the hearts of those that live here.

Like Colonus, the space where these plays were performed in Athens at the Festival of the Greater Dionysia was a kind of sacred space. No one knows exactly how it was viewed by the citizens of Athens. It was clearly not a place, like a temple, where rituals were enacted; but neither was it the completely desacralised space of our modern theatres. The theatre stood within the sanctuary of the god Dionysus, at the foot of the Acropolis, and the front row of seats consisted of marble chairs reserved for magistrates and priests, the central chair being that of the priest of Dionysus. The theatre was used not only for dramatic spectacles but for ceremonies of all kinds and even for meetings of the Assembly. Thus the plays performed there must have had a sacred and ritualistic quality which even the best modern productions cannot hope to emulate. Sophocles has thus elided two spaces, neither of which is 'famous in legend', as was the site of Apollo's shrine in Delphi, say, or even that of the Battle of Marathon, but both of which live in the hearts of those familiar with them: his native Colonus and the space in which the play is being enacted. Both are, 'here, where we are . . .'.

The Lesson of the Hand (2)

Oedipus knows this is where the gods have meant him to come for his final act, though neither he nor the audience yet know what form this will take. Now he begins to feel out by touch that space which the audience has so often looked at and with which it must think that it is thoroughly familiar. The effect of this, as with the climactic scene in the Chaplin film, is to make us experience it for the first time and to make us feel with our bodies the mystery of theatrical representation, where a space is always two things at once (except in *Waiting for Godot* perhaps):

OEDIPUS	Give me your hand.
ANTIGONE	Here, father . . .
OEDIPUS	Further yet?
CHORUS	Further.
OEDIPUS	Again?
CHORUS	Lady, lead him; you understand us.
ANTIGONE	Feel your dark way as I lead you, father . . .
CHORUS	Stay now: you need not come beyond that slab or rock.
OEDIPUS	Here?
CHORUS	It is far enough.
OEDIPUS	I may sit?
CHORUS	To your left, there's a jutting ledge, low down.
ANTIGONE	I'll show you, father. Carefully now –
OEDIPUS	O dear!
ANTIGONE	One step at a time. Lean on my arm.

It is a slow and painful progress, but for the audience a strangely releasing one. In an almost ritualistic fashion Oedipus touches the different parts of the stage. And it is not just any man who is doing this, but a man whose terrible destiny has led to his becoming polluted and therefore a sacred presence. We must also, of course, think of the aged Sophocles ritually touching every corner of the stage which has been his life as he prepares to take his leave of it.

But the play has only just begun. Oedipus, we will discover, is merely passing through this space (and we must remember that these plays had no such thing as a 'run'). He is, as it were, merely leaving a trace of his presence here before moving on. But 'on' is precisely where we cannot follow him, for it is not a place and not exactly a state. It is 'where he will go when he has left this place', and what that might mean the play will in due course try to convey to us.

Before that we witness a battle of wills as first Creon and then Oedipus' son Polynices try to persuade and then, when that does not work, to force Oedipus to return to Thebes with them, for an oracle has said that the side in the fratricidal civil war now raging there which holds the sacred body of Oedipus will be victorious. But Oedipus stands firm, and he is protected by Theseus, who rules over Colonus and Athens, and who ensures that Oedipus will do what he wishes and what the gods have told him, that is to wait here in Colonus for their call.

And now, with Creon and Polynices despatched, the call comes at last. Now it is time even for Antigone and Ismene to let go of their father:

> The hand of the God directs me.
> Follow, my children.
> It is my turn now to be your pathfinder,
> As you have been to me. Come. Do not touch me.
> Leave me to find the way to the sacred grove
> Where this land's soil is to enclose my bones.
> This way . . . This way . . . Hermes is leading me,
> And the Queen of the Nether world. This way . . . This
>   way . . .
> Dark day! How long since thou wast light to me!
> Now ends my life for ever . . .

And so he leaves our sight, followed by his new-found protector, Theseus, to whom he had earlier said:

> No, no: I am a man of misery,
> Corrupt with every foulness that exists!
> I cannot let you touch me. No, you shall not!
> No one but those on whom it lies already
> Can bear this heavy load with me. Stay then
> And take my thanks, so. And be kind to me
> Henceforth, as you are now.

54

Sophocles has reserved his greatest coup for the last. As always with this profoundly conservative and deeply innovative writer, what he does is simply to take a basic convention of the Athenian stage to unexpected lengths. Just as we were made to experience the sacred space of Colonus and of the tragic theatre through Oedipus' groping and polluted hand, so now we are made to experience that which this space is ultimately designed to celebrate, the meeting of man and god, not through something we see, not even through the kind of rich Shakespearean evocation of that which we do not see, like Edgar's 'description' of the cliffs of Dover for us and his blind father, but through a triple denial of the act of seeing.

In most Greek tragedies the audience does not witness the death of the hero but listens as a messenger arrives and tells how that death has occurred, a death which he himself has witnessed. In this play the messenger returns, but only to tell us that what he saw was not a man dying but a man shielding his eyes from the sight of another man passing away. 'People of Colonus!' the messenger begins, 'I am here to say that the life of Oedipus is ended. And there is much to tell of all that I saw happen.' And what he saw was this:

When we had gone a little distance, we turned round and looked back. Oedipus was nowhere to be seen; but the King was standing alone holding his hand before his eyes as if he had seen some terrible sight that no one could bear to look upon; and soon we saw him salute heaven and earth with one short prayer.

And in what manner Oedipus passed from this earth, no one can tell. Only Theseus knows. We know he was not destroyed by a thunderbolt from heaven nor tide-wave rising from the sea, for no such thing occurred. Maybe a guiding spirit from the gods took him, or the earth's foundations opened and received him with no pain. Certain it is that he was taken without a pang, without grief or agony – a passing more wonderful than that of any other man.

It has become commonplace in modern productions of these plays, to have the messenger mime the terrible events he is reporting, as though no director could bear to let an actor merely speak to the audience for more than a few seconds without the need for some kind of action on

55

stage 'to hold the audience's attention'. Sophocles blocks off this possibility straightaway, for what the messenger sees and reports is not some frightful or even wonderful event but something so simple that most dramatists would instantly dismiss the possibility of making it the central feature of the climactic scene of a play: a man holding his hand up before his eyes. Of course the messenger too feels the need to embroider, to try to enter the mystery, and he adds that Theseus held up his hand over his eyes 'as if he had seen some terrible sight that no one could bear to look upon'. But this only reinforces the audience's sense that at some point, and rather sooner than later, interpretation has to stop. All we have, after all, is a man standing before us and telling us about another man standing some way away from him, shielding his eyes with his hand, as though to protect himself from something 'not endurable to see [*oud anaschetou blepein*]'.

And yet that event, triply removed from us as it is, is far more moving than any death on stage could ever be. For such a death is always riddled with falsity. That is why when we are presented with a death on stage we tend, while watching, to think about the quality of the acting or the subtlety of the lighting, about anything in fact but what is purportedly taking place before our eyes. This is not because it is too painful but because something in us revolts at the falsity of what we are being asked to witness. Falsity partly because death is so clearly not taking place, since death is unique and irreversible and we know that this one will take place again and again in the same way at the same time on the following night; but also because, even if it were actually taking place, we, who can only sit in our seats and watch, recognise that we could not be adequate to it, that we are not really *there*.

A massacre is an event; death is not. That is why we are not embarrassed – though we will be horrified – at watching news-reel of a massacre, but feel deeply uneasy about watching the last moments of an individual being filmed. Suicide, on the other hand, *is* an event, and therefore inherently dramatic, as Shakespeare sensed when he wrote Othello's great last speech. Death, ordinary death, however, cannot be dramatised, only told, as the Bible tells it: 'Then Jacob died, being old and full of years, and he was buried . . .'.

The way *Oedipus at Colonus* lets go of its protagonist, on the other hand, moves us with a liberating force. We, who have seen the blind Oedipus groping his way about the stage before us, and who, because of his blindness, have found ourselves inhabiting his body as we never do that of Agamemnon or Oedipus himself when still king, now find ourselves parting from him, and so from ourselves, in a way that is both natural and ungraspable. But the point is that we are not being asked to grasp it. Sophocles has found a way of allowing us to give our assent to that which must be because he has protected Oedipus at the end not only from the rapacity of our gaze but also from the workings of our imaginations. The terrible necessity of loss is made bearable, but neither tamed nor falsified, because Sophocles has succeeded in writing a play in which, to adapt a famous remark about *Waiting for Godot*, we are made to recognise how, for each of us, as for Oedipus, everything happens just once.

9 Praesentia

On my first visit to Los Angeles I surprised my hosts – and myself – by asking to be taken down to the sea. I found that, more even than wanting to visit the streets down which Philip Marlowe had walked, or any of the city's great museums, which my hosts were anxious to show me, I wanted to dip my hand in the Pacific. We drove out of town and along the coast in the direction of the Getty Museum. They stopped the car and I got out and went across the dirty beach and bent down where the waves lapped the shore.

The water was warmer than it ever is in the Channel, though perhaps a little colder than it is in the Mediterranean. It tasted, of course, as I found when I put my hand up to my mouth, of salt. I dabbed my forehead and cheeks, took one last look and trudged back to the waiting car.

Why this need to touch? Why had I not been content simply to see the Pacific? (Why indeed should I have wanted even to see it, since I knew very well what it must look like, no different from any other stretch of sea at the edge of a big city?)

I don't know. I only know that, having got to Los Angeles, I had to do more than merely visit the city, more than merely see the ocean. Dipping my hand in it confirmed in some way that I had been there. Less permanent than a photo (for how long does a sensation last?), the act of dipping my hand in the Pacific nevertheless did what no photo could ever do. It confirmed that I had indeed been there, been there 'in person'. Within a very short while, of course, I could recall nothing of that moment, only my sense of myself hurrying across the sand and the feeling of disappointment that the water did not seem in any sense distinctive.

So, it is not that my hand retains the feel now that my conscious memory has blotted out the event. This was no Proustian experience, to be re-activated later, when I would dip my hand into some other sea. And how could it be, for I was there only for a moment and all my actions were much too conscious and willed for them to be capable of any Proustian resurrection? No. All I am sure about is that this was something I felt driven to do and that now I know, in some obscure way, that I have been on the other side of the world and made contact with one of the world's great oceans. I would not, of course, have made any special effort to go all the way there only to dip my hand into the Pacific; as I was there, though, it felt right, even important, that I should do so. Why it should be important is a mystery to me, but there it is.

A friend who recently visited Rome for the first time since adolescence told me he had never been in a city where he had felt so great a desire simply to touch. When I asked him why he simply said: 'I suppose it's because there's such a sense of antiquity about it. Everything there seems to stretch back so far.' But why, I asked him, had he not been content merely to see, why had he felt the need to *touch*? 'I really don't know', he said, and then: 'I suppose touching something confirms its presence.'

Its presence to you, but also your presence to it. The doubleness is crucial.

St John, in his Gospel, tells how, when the resurrected Jesus returned to his disciples,

> Thomas, one of the Twelve, called Didymus, was not with them . . . . The other disciples therefore said unto him, We have seen the Lord. But he said unto them, Except I shall see in his hands the print of the nails and put my finger into the print of the nails, and thrust my hand into his side, I will not believe. And after eight days again his disciples were within, and Thomas with them: then came Jesus, the doors being

shut, and stood in their midst, and said, Peace be unto you. Then saith he to Thomas, Reach hither thy finger, and behold my hands; and reach hither thy hand, and thrust it into my side: and be not faithless, but believing. And Thomas answered, and said unto him, My Lord and my God. Jesus saith unto him, Thomas, because thou hast seen me, thou hast believed; blessed are they that have not seen, and yet have believed. (John 20:24–9)

Doubting Thomas, he has since been called. What he wants is to confirm for himself that it is indeed the Jesus who was crucified who is standing there before him, and only by touching Jesus' wounds will his doubts be resolved. His attitude to the thing touched, though, is quite different from mine that day in Los Angeles or my friend's in Rome, or indeed from that of the countless pilgrims throughout the ages who have visited the Holy City to touch the foot of the statue of St Peter. (That foot, so worn away by the touch of pilgrims over the centuries, was the one object my friend told me he felt no need to touch. Not that he felt it was any less ancient than the other monuments, but the sight of it already being touched by hordes of tourists with cameras slung round their necks put him off. The object had lost its aura.)

But Jesus' response too, in the Gospel account, drives too sharp a wedge between faith and the world. The desire that drove thousands of pilgrims to the shrines of holy men, as before that to the shrines of local heroes in Greece and Asia Minor, was quite different from the desire of Doubting Thomas for confirmation, and quite other than the failure of faith that Jesus imputes to Doubting Thomas here. Partly, as Peter Brown has beautifully shown, it was the desire to converse with a human intercessor: 'The *praesentia* on which such heady enthusiasm focussed', writes Brown in *The Cult of the Saints*, 'was the presence of an invisible person. The devotees who flocked out of Rome to the shrine of St. Lawrence, to ask for his favour or to place their dead near his grave, were not merely going to a place; they were going to a place to meet a person – *ad dominum Laurentium*.' And he goes on:

The fullness of the invisible person could be present at a mere fragment of his physical remains, and even at objects, such as the *brandea* [little cloths which the pilgrims lowered on to the tomb below and then drew up] of Saint Peter, that had merely made contact with these remains. As a result, the Christian world came to be covered with tiny fragments of original relics and with 'contact relics' held, as in the case of Saint Peter, to be as full of his *praesentia* as any physical remains.

Thomas was, in effect, a modern sceptic, unwilling to trust anything he could not himself see and touch; the pilgrims on the other hand never doubted that virtue lay in the objects that were the goal of their journey. Yet the crucial notion of 'going to a place to meet a person', which Brown is at pains to place at the forefront of his discussion of early pilgrimage, makes it clear that the site of pilgrimage was not just the site of holiness and power, but rather the site of an *encounter*. The practice of pilgrimage has too often, in our post-Reformation world, been dismissed as but one aspect of an essentially 'magical' world-view, common to all primitive peoples, but which we have, fortunately, grown out of. Recently, however, historians like Peter Brown, Richard Southern and Eamon Duffy, and anthropologists like Clifford Geertz and Mary Douglas, have begun to unpack the word 'magical' and, by so doing, have shown up the nature of our own presuppositions while helping us to understand a little more about the workings of cultures not our own. At the same time what they have shown is that we should beware of making too sharp a distinction between cultures – what is 'not our own' may turn out to have more in common with what we do and think than we first imagined. Does Brown's concept of *praesentia*, after all, not help me to understand my own confused feelings that sunny day in Los Angeles?

True, I did not go down to the sea to meet a person, but I did not dip my hand in the water to test that it was really sea water either. Despite the enormous differences between the two experiences, Brown's discussion of early medieval pilgrimage may help us understand the nature of the mysterious impulse we all have to touch the world around us. But we need to follow him with care, and to start by distinguishing the matters he is dealing with from other, rather different attitudes to touching which we know to have existed in the Middle Ages.

10   The King's Touch

When Jesus began to perform his miracles of healing,

> a certain woman, which had an issue of blood twelve
> years, And had suffered many things of many physicians,
> and had spent all that she had, and was nothing
> bettered, but rather grew worse, When she had heard of
> Jesus, came in the press behind, and touched his
> garment. For she said, If I may touch but his clothes, I
> shall be whole. And straightway the fountain of her
> blood was dried up; and she felt in her body that she
> was healed of that plague. And Jesus, immediately
> knowing in himself that virtue had gone out of him,
> turned him about in the press, and said, Who touched
> my clothes? And his disciples said unto him, Thou seest
> the multitude thronging thee, and sayest thou, Who
> touched me? And he looked round about to see her that
> had done this thing. But the woman fearing and
> trembling, knowing what was done in her, came and fell
> down before him, and told him all the truth. And he
> said unto her, Daughter, thy faith hath made thee whole;
> go in peace, and be whole of thy plague. (Mark 5:
> 25–34)

The notion of a healing power emanating from a
holy man is to be found in most societies. So-called faith-

healers still abound in the West today. It is rare, however, to find it expressed quite so directly and powerfully as it is here. More normal is the healing by laying-on of hands and the uttering of a sacred formula, and indeed we find in the Gospels numerous accounts of Jesus touching the affected parts of the blind and the lame, giving them verbal encouragement and sending them on their way whole again. What is perhaps surprising is to discover that the faith in the touch of a sacred person as a cure for disease persisted in the West right through to the beginning of the eighteenth century, though by that time it was no longer the martyr or the bishop who was believed to have the power to heal, but the monarch.

The English reader is likely to have come across the phenomenon first in Shakespeare. In the third scene of Act IV of *Macbeth* Shakespeare shifts the action from Scotland to the English court of Edward the Confessor, and to a conversation between two of Edward's courtiers. One of them says:

> There are a crew of wretched souls
> That stay his cure. Their malady convinces
> The great assay of art; but at his touch,
> Such sanctity hath heaven given his hand,
> They presently amend.
>
> (141–5)

Most annotated editions of *Macbeth* will at this point inform one that the courtier is here talking about the king's evil or scrofula, which, it was believed, could be cured by the touch of the kings of France and England. They usually go on to inform one that the last well-known person to have been touched in this way was Samuel Johnson who, as a three-year-old, was brought to London to be touched by Queen Anne in 1711.

Marc Bloch, the great Jewish-French historian murdered by the Nazis, devoted a whole book to the phenomenon, *Les Rois thaumaturges* (1923). Bloch demonstrates here with his customary erudition and elegance how the idea first arose in France and England just at the moment when it had become vital to strengthen the legitimacy of the ruling dynasties, and how it was reinforced over the centuries by the need to establish the monarch as the equal in every respect of the religious leader, the bishop or archbishop.

The king, writes a courtier at the French court in 1493, 'heals the sick simply by the contact of his hand', and in the description of a stained-glass window in Mont S. Michel we read: 'He touches them one after the other with his right hand, from forehead to chin and from one cheek to the other.' The making of the sign of the cross was not, however, essential; what was essential was the laying-on of hands, for it is the touch of the hands that cures the disease: *ad solum manus tactum certos infirmos sanare dicuntur.*

Bloch shows how every time a new claim was made for the healing powers of the French king the English were quick to up their own claims. His book is thus essentially a study in medieval politics. It reveals the extent and power of the belief with which it deals but it does not really try to explain them. What it does make clear, however, is that the notion of a healing power emanating from a secular or religious leader is quite different from the complex set of expectations set up by the idea of pilgrimage. For though the healing of bodily ailments is obviously frequently one of the motives for pilgrimage, and is perhaps in the West today the only one, it was never the primary one. That, as Peter Brown has shown, was always the desire to go to a place to enter a holy presence. This is also central to the Hindu pilgrimage tradition, where it is called *darshan*: 'To have the *darśana* of a saintly man, a temple, a holy river, an image of the deity, etc.', writes the anthropologist E. A. Morinis, 'is to acquire merit by virtue of simply having been in the presence of some form of radiance of the holy. This concept is important in all Hindu pilgrimage traditions as a sufficient motive for a pilgrim's pilgrimage.' Thus it is in relation to pilgrimage rather than through the healing touch of a sacred being that we can best come to understand what lies behind our own modern and purely secular and instinctive need to touch whenever we travel to foreign lands.

11   The Therapy of Distance

To begin with we must understand what the displacement which pilgrimage entails did for the individual who undertook it. Peter Brown picks up a phrase used by Alphonse Dupront, who describes pilgrimage as 'une thérapie par l'espace', and comments: 'The pilgrim committed himself or herself to the "therapy of distance" by recognising that what he or she wished for was not to be had in the immediate environment. Distance could symbolise needs unsatisfied, so that, as Dupront continues, "le pèlerinage demeure essentiellement départ"; pilgrimage remains essentially the act of leaving.'

This is what Chaucer evidently came to understand in the process of writing *The Canterbury Tales*. Originally he had planned the work so that his pilgrims would tell two tales on the way to St Thomas' shrine in Canterbury and two on the way back. That is the plan as announced in the General Prologue. But though the work as we have it is unfinished, we can see that in the course of writing he changed his mind and decided that he would deal only with the journey *to* Canterbury. Moreover – and this was one of his strokes of genius – he would end with the pilgrims still outside Canterbury, though within sight of it. The pilgrims are still shown going to Canterbury 'the hooly blisful martir for to seke' (I.17 – note the implication here that he is a living presence, not long dead), and there is no sense by the time the Parson speaks his prologue within sight of Canterbury that their journey has been cut short. On the contrary, the Parson says:

> I wol yow telle a myrie tale in prose
> To knytte up all this feeste, and make an ende.
> And Jhesu, for his grace, wit me sende
> To shewe yow the wey, in this viage,
> Of thilke parfit glorious pilgrymage
> That highte Jerusalem celestial.
>
> (X.46–51)

The act of pilgrimage rather than its goal is the focus of the work as we now have it, thereby enabling Chaucer not only to use his fiction as a metaphor for human life and human imagining, but also to remain true to the dynamics of pilgrimage as developed in early Christianity. For, as Brown goes on to show, if 'the distance is there to be overcome', if 'the experience of pilgrimage activates a yearning for intimate closeness', that overcoming is never achieved, that intimate closeness is always deferred. For the pilgrims who arrived after their long journeys

> found themselves subjected to the same therapy [of distance] by the nature of the shrine itself. The effect of 'inverted magnitudes' sharpened the sense of distance and yearning by playing out the long delays of pilgrimage in miniature. For the art of the shrine in late antiquity is an art of closed surfaces. Behind these surfaces, the holy lay either totally hidden or glimpsed through narrow apertures. The opacity of the surfaces heightened an awareness of the ultimate unattainability in this life of the person they had travelled over such wide spaces to touch.

In his account of the pilgrim centre of Tarapith in West Bengal, Morinis, for his part, explains:

> The image which the pilgrim sees when he is led into the temple sanctum is about three feet tall, made of metal and heavily covered in clothing. The face and posture are representative of the Tārā known in conventional iconography – four-armed, tongue lolling, with a garland

of skulls, etc. But it is emphasised by the priests of the temple that this is not the true image of Tārā. Every evening, following the *sandhyā ārati*, pilgrims are allowed to have a *darśana* of the original image. The conventional metal statue, is, in fact, hollow and open at the back.

Pilgrims who wish to see this image queue up and are admitted into the sacred chamber in small groups, but even here there are still boundaries, for 'the stone image of Tārā has no discernible features. The events which are supposed to be depicted there are not easily made out.' The priest unwraps the *sārī* which covers the stone image and then 'points out to the pilgrims the significance of the various protrusions, depressions and other irregularities of the stone'. And Morinis concludes: 'Nīlakaṇṭha and Devī would not be evident there to the unguided eye.'

Morinis explains how the pilgrim, once he has arrived at Tarapith, makes his way into the temple and to the sacred chamber. Brown, for his part, argues that the pilgrim journey was in a sense reduplicated in the actual sites:

> In Tebessa, the approach to the shrine, as it wound past high walls, swung under arches, crossed courtyards, and finally descended into a small half-submerged chamber, was a microcosm of the long journey of pilgrimage itself. At the shrine of Saint Lawrence at Rome, the first sign of patronage by a Christian emperor involved a heightening of the effect of distance; Constantine installed flights of stairs leading up and down from the grave and 'shut off' the grave itself with a grille of solid silver weighing a thousand pounds, thus keeping the tomb of Lawrence still at a short distance from the pilgrims.

At the shrine of St Peter, what Brown calls 'a whole ritual of access' was played out, for, in the words of Gregory of Tours which Brown cites: 'Whoever wishes to pray here must unlock the gates which encircle the spot, pass to where he is above the grave, and, opening a little window, push his head through and there make the supplication that he needs.' There were little golden keys with which to open the gates, and these were of course treasured by those who made the pilgrimage to Rome as much as the *brandea* which, as Brown explains, were 'heavy with the blessing of Saint Peter'.

Of course this ritualisation of the tension between proximity and distance sometimes broke down, but the occurrences of such instances only brings out more clearly the underlying principle of the therapy of distance. Brown describes, for instance, how a Carthaginian noblewoman, Megetia, 'committed herself to the "therapy of distance" by travelling away from her family to the shrine of Saint Stephen in nearby Uzalis'. But, once there, matters took a course of their own. Her early biographer describes how,

> while she prayed at the place of the holy relic shrine, she beat against it, not only with the longings of her heart, but with her whole body so that the little grille in front of the relic opened at the impact; and she, taking the Kingdom of Heaven by storm, pushed her head inside and laid it on the holy relics resting there, drenching them with her tears.

And Brown concludes: 'The carefully sustained tension between distance and proximity ensured one thing: *praesentia*, the physical presence of the holy . . . was the greatest blessing that a late-antique Christian could enjoy.' Morinis' detailed examination of what goes on in three West Bengali shrines to this day would suggest that it was central not only to the life of late-antique Christians.

The therapy of distance induced by pilgrimage was a double one. The pilgrim set out, left the safety of the known and the familiar, and engaged in what Victor Turner has called a 'liminal' or 'liminoid' experience, a rite of passage. But this was not so much a passage from A to B as a journey into the experience of distance itself. When the pilgrim touched the shrine at the end of his or her long journey it was in an attempt not so much to bridge the distance that separated him or her from the holy as an instinctive way of making that distance palpable. There was no need for more, no need to clasp or grasp, no need to try and make the saint one's own or to try and prise the 'virtue' out of him. (That, at least, was the theory. In practice, as we see with Megetia, we human beings always want more.) When Thomas sought to thrust his hand into Christ's wounds it was to lay a doubt to rest; when the woman touched Jesus' garments it was to cure her ailment by contact with the source of all health; but when the pilgrim travelled to the shrine of St Lawrence or St Thomas it was in the confidence that the act of pilgrimage itself would make the saint look kindly on him when he arrived. Having come all that way to 'seek' the saint he would feel that his presence in this place had brought the distance between them to life and this, *darśana, praesentia*, was enough.

By the seventeenth century that confidence had evaporated and distance had become the enemy. Descartes is determined to get down to what can be known certainly by being known *immediately*, without mediation. Berkeley, in a horrifying image, wants reality (or God, since to him God is reality-as-we-perceive-it) to be *as near as pain*: 'A man born blind', he writes, 'being made to see, would, at first, have no idea of distance by sight . . .. The objects intermitted by sights would seem to him (as in truth they are) no other than a new set of thoughts or sensations, each whereof is as near to him as the perceptions of pain or pleasure, or the inward passions of the soul.'

We are all heirs of the seventeenth century. We all still talk quite naturally of getting down to the truth, of clearing away the clouds of confusion, as if the truth lay buried beneath obfuscating material which only needed to be removed in order for it to shine forth. But it may be that the assumptions which led to the development of the medieval practice of pilgrimage, which informs Hindu pilgrimage to this day, are not simply cultural and not simply to be washed away by changes in philosophical and popular attitudes. It may be that they are more deeply rooted in the body's relation to the world than we care to acknowledge and that the Middle Ages and Hindu culture have merely been more receptive to our basic needs. For if the goal of pilgrimage was, and is, to make distance palpable, is that not still an instinct with us today? Is it not what we have found Coleridge discovering in his lime-tree bower? Is not the secular visitor to Rome or Los Angeles, who touches the age-old stones or the ocean, instinctively repeating that ancient gesture and, by so doing, not so much bridging the distance between himself and these repositories of power as acknowledging their otherness and the awe he feels in their presence?

But how can that be? For the medieval pilgrim the power emanated from the saint. If we no longer believe in saints do we simply transfer that power to stone monuments or to natural phenomena like the Pacific Ocean or Victoria Falls? That is clearly not quite right. Neither I in Los Angeles nor my friend in Rome was suddenly overcome by some form of animism. I think the awe we both felt – and felt we needed to acknowledge by the act of touching – had to do rather with the unimaginable size and ancientness of the ocean, the sheer length of time through which those stones had stood in Rome. By touching, I think, we experience a sense of our own implication in a history longer and broader than our personal one: I am – and it is – and touch can somehow affirm that truth. This, I think, is the implication of that act of touch.

As it happens, Proust can help us once again. At the end of his great essay on reading, which he wrote as an introduction to his translation of Ruskin's *Sesame and Lilies*, he arrives at a contemplation of the very mystery I have been exploring. He has been talking about the fact that it is as much in the silence between words as in the words themselves that we find what moves us in the ancient books we love. In particular he has been pondering why it is he is moved by the colons which divide the periods in the Gospels almost more than he is moved by the actual content of those periods, and it is the image of the two dots, I suspect, which leads him on to his conclusion. The classics, he says, move us not only because they have profound things to say and say them in moving and beautiful language, but because they bring into the present – our reading time – a portion of the past. 'How many times', he says,

in *The Divine Comedy*, in Shakespeare, have I known that impression of having before me, inserted into the present actual hour, a little of the past, that dreamlike impression which one experiences in Venice on the Piazzetta, before its two columns of gray and pink granite, that support on their Greek capitals, one the Lion of Saint Mark, the other Saint Theodore trampling the crocodile under his feet – beautiful strangers come from the Orient over the sea at which they gaze in the distance and which comes to die at their feet, and who both, without understanding the conversations going on around them in a language which is not of their country, on this public square where their heed-less smile still shines, keep on prolonging in our midst their days of the twelfth century, which they interpose in our today [elles continuent à attarder leurs jours du xiie siècle dans la foule d'aujourd'hui, sur cette place publique où brille encore distraitement, tout près, leur sourire lointain].

What seems to fascinate Proust is that these two columns, having been wrenched from their temporal and geographical origins, continue to exist as if nothing had happened, as if unaware of anachronism or estrangement, oblivious of their colon-like disjunctive interposition. They keep their distance, like the Virgin of Amiens, precisely because they refuse to acknowledge it. 'Around the pink columns', Proust goes on, 'surging up toward their wide capitals, the days of the present crowd and buzz. But, interposed between them, the columns push them aside, reserving with all their slender impenetrability the inviolate place of the Past . . ..' And it need not be stones that convey this experience. Over there, I think, is that other ocean, and the memory of my action that day in Los Angeles (rather than the memory of the actual sensation) helps to establish its distance from me now, its otherness, its incomprehensible vastness and ancientness, and so makes real for me its presence in a world which I also inhabit, and helps confirm that it is and will remain quite other than all my imaginings.

12   The Therapy of Distance (2)

'The opacity of the surfaces', writes Peter Brown,
'heightened awareness of the ultimate unattainability in
this life of the person they had travelled over such wide
spaces to touch.' The lay-out of the pilgrimage centre and
the shrine was designed to heighten the effect of distance
while reinforcing the message that distance may be benign,
that it does not have to be totally overcome for *praesentia*
to manifest itself. What we find here, as often in earlier
and more traditional cultures, is the ritualisation of a basic
and inescapable fact of life, the giving to it of a form and a
shape in order to make it bearable and comprehensible.
The ritual of pilgrimage, as both Brown and Morinis
describe it, seems to enact in an atmosphere open to the
imposition of public meanings what every child
experiences when he holds his mother's hand. The mistake
of the Lady Megetia was merely the mistake made by every
insecure child, that of turning touch into grasp, of trying
to grip and possess what should only be lightly touched or
simply held.

But the late medieval church authorities, fearful that
pilgrimage was being misused, that too many people were
going on pilgrimages in the mistaken belief that contact
with the saint would cure them of their physical ailments
or else merely to satisfy their curiosity or engage in
amorous play far from the prying eyes of their spouses or

neighbours, tried to tighten their control over pilgrimages and even to restrict or ban them altogether. The proto-Protestant adherents of the *Devotio Moderna* even went so far as to insist that one should substitute mental pilgrimages for the real thing. One should stay at home, they argued, and merely *meditate* on the cult image. Interestingly enough, this too has its counterpart in the Hindu and Moslem traditions. Morinis quotes the poet-theologian Kabir as saying: 'Benares is to the East, Mecca to the West; but explore your own heart, for there are both Rama and Allah.'

Craig Harbison, in his book on van Eyck, points out that 'the kind of mental pilgrimage promoted by the Modern Devotion was a lengthy and methodical affair. This was not an imagined quick trip to a familiar local shrine, but a long, arduous mental journey to the Holy Land, tracing the steps of Christ, especially in order to relive the Passion.' And he suggests that some of van Eyck's paintings, as well as those of other Netherlands artists of the fifteenth century, were precisely designed to promote such mental pilgrimages. A similar argument has been put forward by Donald Howard in relation to *The Canterbury Tales*.

But although such notions are superficially attractive they are, I think, profoundly misleading. For what the reformers wanted was to control what seemed to them to be something unruly and therefore irreligious; what they failed to see was that by cutting out the abuse they would also be impoverishing the very springs of religion itself. But that is a familiar story in the history of religion.

There is no substitute for the therapy of distance. To abolish the actual journey, with all its attendant dangers and attractions, with all its temptations and seductions, was also to abolish the therapy. As Eamon Duffy, writing about late medieval pilgrimage, with all its accretions of magic, bigotry and superstition tellingly puts it:

Even for the healthy pilgrim, drawn to the shrine by an itch for travel, by simple devotion, or by the desire to obtain indulgences, the ranks of crutches and fetters, boats and legs and hearts, and the sight of the sick themselves, waiting with varying degrees of hope and impatience, were an assurance that God was in his Heaven and the Devil did not always have the last word.

He cites a short doggerel prayer to the fictional Saint Walstan of Bawburgh, whose cult had grown prodigiously in Norfolk between the late thirteenth and the early sixteenth centuries, and whose shrine was the focus of an annual pilgrimage on his feast-day, 30 May:

> You knight of Christ, Walston holy,
> our cry to thee meekly we pray;
> Shield us from mischiefe, sorrow & folly,
> engendring and renewing from day to day,
> replenishd with misery, Job doth truly say,
> & bring us to health blessed with [Jesus'] right hand
> him to love & know in everlasting land.

'One could have worse summaries of the objectives of late medieval religion,' Duffy comments, 'and of the place of the cult of saints within it.'

By contrast the devout follower of the precepts of the Modern Devotion, who experienced pilgrimage only in his head, in the privacy of his study, through silent meditation, was already well on his way to becoming the prisoner of solitary confinement, the isolated individual prey to the *anomie* of modern urban life, cut off from God and his fellows and thus ultimately from himself, whom we have all encountered, if not in the mirror, at least in the writings of the classical sociologists.

A fascinating Lollard text has come down to us, which purports to be the record of the examination of a certain William Thorpe by Archbishop Arundel some time in the late 1390s, that is, just when Chaucer was writing *The Canterbury Tales*. Thorpe argues that there are two kinds of pilgrimage, the true and the false. True is 'pilgrims travelling towards the bliss of heaven' by living virtuous lives; but most of those who now go on pilgrimage, Thorpe says, are not like that at all – 'these same pilgrims, telling the cause why that many men and women now go hither and thither on pilgrimage, it is more for the help of their bodies than for the help of their souls'. They sing and play the bagpipes, it seems,

> so that in each town that they come through, what with noise of their singing, and with the sound of their piping, and with the gingling of their Canterbury bells, and with the barking out of dogs after them, these make more noise than if the king came there away with his clarions and many other minstrels. And if these men and women be a month out in their pilgrimage, many of them an half year after should be great janglers, tale tellers and liars.

The archbishop pulls him up at this and reminds him that music helps the sore and tired to keep going. Besides, he asks, what about King David? Did he not play the harp and dance before the ark of the Lord? Thorpe's only reply to this is that we should read not the letter but the spirit of Scripture, and when Arundel asks him whether it is not right to have organs in church wherewith to praise God, he replies: 'Yea, sir, by man's ordinance, but by the ordinance of God a good sermon to the people's understanding were . . . much more pleasing to God.'

These two minds will never meet. Thorpe focuses on the temptations to sin afforded by pilgrimage, poetry, music and song, while Arundel replies with the traditional defence of these as an aid to religion. Each is convinced that he has won the argument, but today we can see that the questions have been wrongly posed. It is not a matter of letter versus spirit, true pilgrimage versus false. There can never be a simple distinction between the two: pilgrimage, like poetry and music and storytelling, is, for better or worse, embroiled in both or it is nothing at all.

Few of us any longer go on pilgrimages but we still respond to works of art, though the old debates about spirit and letter still bedevil aesthetics, even though the terms have changed. It is clear that the work of art, whether it be a painting or a poem, if it is actively engaged with, forces the viewer or reader to undergo a journey no less open to the pitfalls awaiting the unwary or the easily tempted than the real journey. For it too is, in the end, subject to no control once the process of viewing or reading has begun. And Chaucer for one was well aware of this. *The Canterbury Tales* is not a substitute for but an analogy to pilgrimage. Chaucer's Parson, like Thorpe the Lollard, may wish to 'knit up all this mateere' at the end by delivering a sermon which will put the other tales in their place once and for all, but he cannot succeed. His sermon may come at the physical end of *The Canterbury Tales*, but he cannot close the *Tales* off. The reader has been made to feel, long before we get to the Parson's Tale, that there is no end, that the words of the Pardoner are no less (and of course no more) authoritative than those of the Parson, those of the Miller as worthy of our attention (though not more so) as those of the Knight. Going on a pilgrimage, for Chaucer, means having to encounter the Miller and the Pardoner, the Shipman and the Summoner, as well as the Knight, the Man of Law and the Parson. It means having to make up our own minds about the tales of the Merchant and the

Nun's Priest, and that means being able to change our minds, it means having to respond to all the contradictions inherent in the Prioress and her anti-Semitic tale and in the Wife of Bath and her poignant and confused confessions. Indeed, the 'therapy of distance' in *The Canterbury Tales* begins to work for us as soon as we start to read the work, and it is not radically altered by the arrival of the pilgrims within sight of Canterbury and the shrine of the 'hooly blessed martir' they have come to seek. With this book, as with pilgrimage itself, we can touch the mystery but never make it wholly our own.

The Reformation brought about the effective end of pilgrimage in northern Europe, as it did of the processions through the streets of the great cities on the feast-days of the Church and during the presentation of the Corpus Christi cycles which had, since the fourteenth century, celebrated communally and out in the open the common history of all Christians. The corporation records at York apprise us of the decision in May 1561 that 'for as moche as the late fest of Corpus Christi is not nowe celebrated and kept Holy day as was accustomed it is therefore agreed that on Corpus Christi even my lord Mayor and aldermen shall in makyng the proclamation accustomed goe about in semely sadd apparell and not in skartlet', a provision, says Eamon Duffy, who quotes the passage, which was repeated in many other towns. 'By the end of the century in most communities the plays were no more than a memory,' he writes,

> and, though the young Shakespeare may have witnessed one of the last performances of the Coventry Corpus Christi cycle, which survived into the mid-1570s, only the older members of the audience of *Hamlet* would have known at first hand what 'out-Heroding Herod' actually involved. Two centuries of religious drama, and a whole chapter in lay appropriation of traditional religious teaching and devotion, were at an end.

Instead of taking part in public events Christians were now urged to meditate inwardly on Christ's Passion (while, to satisfy their curiosity they could read accounts in cheap pamphlets of the recent voyages of discovery and of the mores of the criminal classes). In such a climate it was only in the realm of art that the 'therapy of distance' was kept alive – in the plays of Shakespeare himself of course and his fellow-dramatists, but also in the quieter realm of lyric poetry.

It has often been remarked how powerful was the influence of St Ignatius' meditational exercises on Donne and his followers. These exercises were in fact elaborations and codifications of precisely that 'inner pilgrimage' which the *Devotio Moderna* had encouraged in the fifteenth century. That they were among the most powerful instruments of the post-Tridentine Catholic Church should come as no surprise, since one of the key facts about the religious controversies of the fifteenth and sixteenth centuries is that the two sides held many assumptions in common and that both had, in effect, lost touch with the attitudes that had held sway in Europe from late antiquity to the fifteenth century.

But while there is no doubt that Donne, the one-time Catholic and descendant of the greatest English martyr since Becket, Thomas More, was profoundly influenced by Ignatian meditational procedures, his poems can never be reduced to these. And he knows this, deplores it, and at the same time relishes it. In the great poem he wrote when he was gravely ill he plays and puns on his own name with a mixture of horror and delight that is itself an index of the irreducibility of poetry and of the acceptance of its provisional, uncontrollable nature:

> Wilt thou forgive that sin where I begun,
>> Which was my sin, though it were done before?
> Wilt thou forgive that sin, through which I run,
>> And do run still: though still I do deplore?
>>> When thou has done, thou has not done,
>>>> For, I have more.
>
> Wilt thou forgive that sin which I have won
>> Others to sin? and, made my sin their door?
> Wilt thou forgive that sin which I did shun
>> A year, or two: but wallowed in, a score?
>>> When thou hast done, thou hast not done,
>>>> For, I have more.
>
> I have a sin of fear, that when I have spun
>> My last thread, I shall perish on the shore;
> But swear by thy self, that at my death thy son
>> Shall shine as he shines now, and heretofore;
>>> And, having done that, thou hast done,
>>>> I fear no more.

The ringing conclusion suggests triumphant closure, but that is only part of the truth. For, syntactically, the last sentence, which is in effect the last stanza, remains completely open. I fear, Donne says, that in the end I will not be saved; so swear, he implores God, that this will not be so, that you will, through your Son, save me; having sworn this, he concludes, you have done all I could ask of you and my fears will vanish. But the lines leave it open as to whether God has indeed acceded to Donne's request before the start of the last two lines or whether that remains only the fervent desire of the sick poet, who will never, till the actual moment comes, know whether or not his desire will be fulfilled. As so often with Donne it is not the clever paradoxes that make the poem but the profound ambiguities at its heart.

Distance is here both therapy and anguish, and however often we read the poem we will never be able to say which triumphs. There is always more. One is never done. Nor is Donne, ever, fully, Donne or done. To that extent Kierkegaard and Derrida are right, and those, from the Lollards to the structuralists, who would seek to eliminate that supplement, are wrong. It was the genius of the early Church and, no doubt, of other religious traditions, to ritualise this in the experience of pilgrimage to the sites of burial of the saints and holy martyrs. When, for historical reasons, pilgrimage came to an end in western Europe, it was left to artists to take up the torch.

13   Relics

'A hectic trade in, accompanied by frequent thefts of,
relics, is among the most dramatic, not to say picaresque
aspects of Western Christendom in the Middle Ages',
writes Peter Brown. The reason for this was that it was
easier to move bits of the bodies and clothing of holy men
than to get large numbers of people to go *to* them:
'Translations – the movements of relics to people – and
not pilgrimages – the movement of people to relics – hold
the centre of the stage in late-antique and early medieval
piety.' But again Brown is concerned lest we accept too
readily the dismissive attitude to relics of the Reformation
polemicists, and anxious to bring back to our awareness
the subtle and complex role which relics played in the life
of the people of Europe for close on fifteen centuries.

Though they might be translated from their relatively
inaccessible place of origin to one within easy reach of a
particular community, relics came bringing with them a
sense of their place of origin. And no one forgot that it was
God who had given the relic, first by letting it be found
and then by allowing it to be moved. Thus St Augustine
comments on the miracles surrounding St Stephen, the
first Christian martyr: 'His body lay hidden for so long a
time. It came forth when God wished it. It has brought
light to all lands, it has performed such miracles.' And
Brown concludes: 'The discovery of the relic, therefore,

was far more than an act of pious archeology, and its transfer far more than a strange new form of Christian connoisseurship; both actions made plain, at a particular time and place, the immensity of God's mercy . . . They brought a sense of deliverance and pardon into the present.'

What this suggests is that even the translation of relics did not abolish the 'therapy of distance'. The discovery, translation and installation of relics made concrete the mercy of God. It was not the relic itself that was important, as Brown says, so much as 'the invisible gesture of God's forgiveness which had made it available in the first place; and so its power in the community was very much the condensation of the determination of that community to believe that it had been judged by God to have deserved the *praesentia* of the saint'. In other words, the precise events of the discovery of the relic and the ceremonies with which it was brought and installed counted for much more than the mere fact of its presence in a particular community.

The relic and the ceremonies associated with it were of equal importance and could no more be separated than the crossing of the Red Sea by the Israelites or the Passion of Jesus could be separated from the celebration of those events amongst Jews and Christians. In both instances the power of the stories and rituals depended on a double focus: God had been merciful, but terrible things had also taken place, and neither aspect must be forgotten. As Brown puts it:

While the relic might be discovered, transferred, installed, and the annual memory of the saint be celebrated in an atmosphere of high ceremony associated with unambiguously good happenings, the relic itself still carried with it the dark shadows of its origin; the invisible person, whose *praesentia* in the midst of the Christian community was not a token of the unalloyed mercy of God, had not only once died an evil death; but this evil death had been inflicted by an evil act of power. The martyrs had been executed by their persecutors . . . Their deaths, therefore, involved more than a triumph over physical pain; they were vibrant also with the memory of a dialogue with and a triumph over unjust power.

By the sixteenth century all this had been forgotten, at least by those in authority. In 1535 Thomas Cromwell sent out his men to the monasteries of Britain on a fact-finding mission, their brief obviously being to provide him with enough ammunition to destroy, not reform them. The general injunction for the visitation, as it was called, stipulated that religious houses 'shall not show no reliques, or feyned miracles, for increase of lucre', and it was not difficult for the visitors to find evidence of such practices. From Bath Abbey, for example, one of them, Richard Layton, wrote: 'I send you vincula S. Petri, which women put about them at the time of their delivery . . . I send you also a great comb called St. Mary Magdalen's comb, and St. Dorothy's and St. Margaret's combs . . .' From Bury St Edmunds John ap Rice reported: 'Amongst the reliques we founde moche vanitie and superstition, as the coles that Saint Laurence was tosted withall, the paring of St. Edmundes naylles, S. Thomas of Canterbury penneknyffe and his bootes, peces of the olie crosse able to make a hole crosse of . . . with suche other . . .'

But how one reads the evidence depends on one's presuppositions. There can be no doubt that for Cranmer and his men these were evident signs of the gullibility of the common people and their exploitation by a rapacious Church. But, as Duffy points out:

> Everywhere one turns in the . . . records of the visitation one finds evidence of large-scale resort by the people to the monastic shrines as centres of healing and help . . . . In attacking monastic 'superstition'. . . Cromwell's men were striking at institutions with a central place in popular religious practice, perhaps most unexpectedly in the domestic intimacies of pregnancy and childbirth. In such widespread evidence of the integration of the monastic shrines into the fabric of popular religion, however, the visitors saw, or chose to see, nothing more than evidence of large-scale exploitation of simple believers.

The stripping of the churches and monasteries, the destruction of images and the reformation of the liturgy were all carried out with that mixture of righteousness and rapaciousness which seems to characterise religious conflicts. Though the transition from medieval to modern is particularly clearly displayed in sixteenth-century England, it was of course not confined to that country. Duffy's final words in *The Stripping of the Altars*, though specifically about England, are in fact a dirge on the death of a culture which had flourished in Europe for a millennium (and which went on flourishing in other parts of the world till our own day), and with the consequences of whose passing we are still, whether we realise it or not, trying to come to terms today: 'The price for such accommodation [i.e. the "Elizabethan Settlement"]', writes Duffy,

> was the death of the past it sought to conserve . . . Cranmer's sombrely magnificent prose, read week by week, entered and possessed their minds, and became the fabric of their prayer, the utterance of their most solemn and their most vulnerable moments. And more astringent and strident words entered their minds and hearts too, the polemic of the *Homilies*, of Jewel's *Apology*, of Foxe's *Acts and Monuments*, and of a thousand 'no popery' sermons, a relentless torrent carrying away the landmarks of a thousand years. By the end of the 1570s, whatever the instincts and nostalgia of their seniors, a generation was growing up which had known nothing else, which believed the Pope to be Antichrist, the Mass a mummery, which did not look back to the Catholic past as their own, but another country, another world.

14   The Girdle and the River

The cult of relics in the Middle Ages is inseparable from
the social and religious world in which it was embedded. It
is this which holds in check the natural human propensity
to gather and possess, to keep close to one objects that will
effect instant cures for bodily and spiritual sickness. Of
course the Reformers' antagonism to pilgrimage, to relics
and to the entire cult of the saints in the late Middle Ages
does have a point. Much of it did reflect the greed of the
Church and the gullibility of the people. But what
sympathetic studies like those of Brown for the beginning
and Duffy for the end of the period bring out is that terms
like 'gullibility' are not entirely transparent. No object was
ever thought to be able to effect a cure by simple contact;
everything depends rather on the world picture of the
seeker, and this cannot be separated from that of the
society as a whole. Where plague, famine, oppressive lords
and infant mortality were rife, how important to feel that
the world was not totally evil or incomprehensible, that the
saints, though dead, were still there to care for you, and
that it was still possible to enter into contact with God
himself through his Son, whose death and resurrection
were celebrated daily in the Mass. The issues are never as
clear-cut as contemporary polemicists or later historians of
ideas like to suggest, which is why poets and painters, who
thrive on ambiguity, usually tell us more about the life of

the past than theologians or lawyers. The author of those remarkable late medieval English poems, *Pearl* and *Sir Gawain and the Green Knight*, for example, was profoundly interested in the subtle conflicts and inevitable overlaps between 'magic' and 'religion', holding and grasping, and he has much to tell us about how men react when their natural trust comes under pressure.

Sir Gawain and the Green Knight hinges on the difference between, on the one hand, a pentangle and an image of Mary inscribed on Gawain's shield, and, on the other, a girdle which, at a moment of crisis, is offered him by a beautiful lady and which, she claims, has the magical property of being able to protect one from death. The poet explains that the pentangle or 'endless knot' signifies, among other things, the five wounds of Christ and the five virtues of 'fraunchyse', 'felawschyp', 'clanness' (i.e. purity), 'cortasye' and 'pite' (pity and piety both). Although it is what Coleridge would call a natural symbol, meaning that we just have to look at it to sense its meaning (five evenly balanced elements held in tension), it has also been 'set' by King Solomon in ancient time, meaning that Solomon has given it a meaning which society has endorsed. The girdle on the other hand belongs to a lady he doesn't know, who gives it to Gawain with the assurance that it possesses life-preserving qualities, though neither its appearance nor any of the founding stories of the society could have alerted us to this beforehand – it is, after all, merely an item of her clothing. The poem turns on which of these objects possesses more power and on the kind of power each possesses.

The narrative opens in ambiguity, as it will go on. Arthur and his court are celebrating Christmas, the dark time of the year made bearable by the birth of the Saviour, one of whose key sayings is that only those who become as little children will enter the Kingdom of Heaven. But if Christmas is the ritual celebration of the child in all of us, the question always remains where to draw the line between the child*like* and the child*ish*, a grey area the poet exploits so as to leave us uncertain about the qualities of Arthur and his court as they playfully celebrate the feast and wait impatiently for some adventure to take place, some unexpected event which will light up the short dark days: are they only waiting to be entertained, waiting for the medieval equivalent of the conjuror to turn up, or are they waiting to witness a miracle akin to Christ's birth, in which His resurrection is already implicit? Can the two kinds of events even be clearly separated?

Enter the Green Knight on a great horse and his challenge to the court: let one of your knights strike off my head here and now, but be prepared to submit to the return blow in a year's time. Gawain takes up the challenge, chops off the knight's head, and then he and the court watch in amazement as the knight picks up the severed head, which now speaks, reminding Gawain of their rendezvous a year hence, and then the knight rides away, still holding his severed head in his hand. Are we witnessing a conjuror or a miracle? And, if the latter, is it a benign analogy of Christ's own miraculous resurrection or its demonic parody?

Gawain arms himself and rides out into the unknown, holding before him his shield with its image of the pentangle and the Mother of God. Will they be enough to protect him when the moment comes, as it assuredly will? The answer given by this subtle and sophisticated poet is not that it will or it won't, but that, when the crunch comes, Gawain fears it won't. For when, with the prospect of a meeting with the Green Knight looming, the seductive wife of his host offers him her girdle, which she says will protect him from death, he succumbs to the temptation, even though it will mean lying to his host. Mary and the pentangle are all very well, but who can be sure that they will offer the same immediate protection? They may be profoundly meaningful symbolically, but what is needed at the moment is not a symbol but effective practical help, and this the lady seems to be offering. So Gawain dons the girdle, saying nothing either to his host (to whom he had promised to return each evening whatever the lady gave him during the day) or to the priest to whom he goes for confession before setting out finally to meet the Green Knight.* This, as both the knight and Arthur, when Gawain eventually returns to court, tell him, was a perfectly understandable and human, though reprehensible, weakness. Gawain, however, once his actions have been revealed, is overwhelmed by guilt and remorse at having succumbed. He cannot face the fact, it seems, that the world will know that he is not the perfect Christian knight everyone took him to be. Even less, we suspect, can he face the fact that he is not as perfect as *he himself* had imagined himself to be. It is going to take a brilliant and subtle act of

---

* The poet is deliberately vague about Gawain's confession, as he is about Arthur's childishness in the first scene; it may be that Gawain does tell the priest about the girdle, but since he makes no effort to restore it to its owner, such a confession is worthless. See John Burrow's excellent discussion in *A Reading of Sir Gawain and the Green Knight*, p. 200.

statesmanship on Arthur's part to correct him and return him once again to the community of his fellows.

How does Arthur do this? Gawain returns to the court, after his encounter with the Green Knight, in which his life has been spared but his neck just 'nicked' once, to chastise him for taking the girdle without saying anything about it, wearing the girdle, now the sign of his sin, as a baldric, visible to all. In a highly emotional manner he recounts the story of his temptation, his fall and his present contrition. In response Arthur merely orders the whole court to wear similar girdles in similar fashion. By this simple act Arthur performs for the girdle what Solomon was said to have done for the pentangle, he 'sets' it, giving it a public meaning. But what kind of a meaning is it? Arthur's action seems to suggest that he wants Gawain to know that he thinks him no more guilty than any of the other courtiers. At the same time, though, his action is a tacit reproof to Gawain, implying that he is no more special than any other member of the court, and that in asserting that he is specially tainted he is committing the sin of despair, which is the mirror image of the sin of pride, for if the proud man sets himself above all others, the despairing man sets himself below all others, asserting in effect that he alone is not worthy of Christ's mercy. But no one, Arthur's action suggests, is beyond such mercy, we are all part of one fellowship, and his tacit injunction to Gawain is therefore akin to Father Zosima's quiet remark to Fyodor Karamazov: 'Do not be so ashamed of yourself, for that is at the root of it all.'

The poet does not labour the point. Indeed, the whole of this rich climactic scene takes up less than twenty lines. Yet with extraordinary sophistication he has uncovered the springs of human narcissism, the ways we lie to ourselves in order to preserve the image we have of ourselves and the ways in which we try to maintain that image even at the cost of extreme self-mortification. For, as Dostoevsky understood so well, self-mortification is usually only the sign of a refusal to face up to what one really is, which is in turn the result of a feeling that there is no one to pardon us.

The poet also shows us what happens when the symbolic systems we live by come under pressure. At the moment of crisis, with his encounter with the Green Knight suddenly imminent, Gawain instinctively opts for the magic girdle, as so many, for example, have opted for collaboration with the enemy when under occupation. The lesson he has to learn is that he should have trusted those symbols of a world in which he professed to believe, and no doubt imagined he believed, the pentangle and Mary's image, and that his real danger arose only when he surrendered that trust. But the poem takes up no moralistic stance, it merely shows the consequences of the action Gawain actually chose, just as Proust merely shows what happens when Marcel tries to hold on to his mother. Neither writer suggests that his protagonist, in the course of living through the dense web that is his life, could have done otherwise, though both recognise that it would have been better for them had they not done what they did. Now, they will have to live with the consequences for the rest of their lives.

In his other major poem, *Pearl*, this extraordinary anonymous fourteenth-century English poet was also concerned with the difficulty of reconciling our system of beliefs with our gut reactions, only this time he deals with it directly rather than through the prism of romance. The protagonist is a bereaved parent who cannot accept, cannot, we would say, come to terms with, the death of his child. He describes her as his 'precious pearl', who, he says, slipped through his fingers one summer's day and fell into the earth where she was lost. The little girl appears to him in a dream and, in an extended debate, tries to make him see that she is now happy and fulfilled in Paradise and that he has no reason to grieve. But though he can grasp what she says intellectually he cannot accept the loss to him her death entails. In the end, desperate to join her on the other side of the river which has separated them throughout their conversation, he rushes forward to seize hold of her. But the shock of his contact with the cold water of the river wakes him up. As with Gawain's painful discovery that the Green Knight had known of his cowardice all along, the shock effects a change in his whole being which mere argument had failed to do. Like Gawain in the face of Arthur's injunction to the court to wear identical girdles to his, the bereaved father is now at last able to let his child go, to understand with his whole being and not just his mind that we cannot ever fully comprehend what happens to us or protect ourselves against the blows of fate. And he understands at the same time that we must learn not merely not to resent this fact but actively to celebrate it. The last stanza of the poem, so reminiscent of George Herbert, has the father partaking of the daily ritual of the Mass, where the bread of life is eaten and where the act of participation implies a recognition that life goes on despite the losses we suffer, and that this is itself a good which cannot be questioned:

Touch

To pay the Prince other sete saghte
Hit is full ethe to the god Krystyin:
For I haf founded hym, bothe day and naghte,
A God, a Lorde, a frende ful fyin.
Over this hyul this lote I laghte,
For pyty of my perle enclyin,
And sythen to God I hit betaghte
In Krystes dere blessyng and myn,
That in the forme of bred and wyn
The preste uus schewes uch a daye.
He get uus to be hys homly hyne
And precious perles unto his pay.

Amen. Amen.

15 'A Goose which Has Grown in Scotland on a Tree'

In 1638 Georg Christoph Stirm, a German student, wrote a letter home from England in which he described what he saw in the famous house-museum of the horticulturalist, John Tradescant. The letter gives a startling insight not only into the contents of Tradescant's museum but also into the mind of the writer who reports on those objects. 'In the museum itself,' writes Stirm,

> we saw a salamander, a chameleon, a pelican, a remora, a lanhado from Africa, a white partridge, a goose which has grown in Scotland on a tree, a flying squirrel, another squirrel like a fish, all kinds of bright coloured birds from India, a number of things changed into stone, all kinds of shells, the hand of a mermaid, the hand of a mummy: a very natural wax hand under glass, all kinds of precious stones, coins, a picture wrought in feathers, a small piece of wood from the cross of Christ . . ., many Turkish and other foreign shoes and boots, a sea-parrot, a toad-fish, an elk's hoof with three claws, a bat as large as a pigeon, a human bone weighing 42lbs., Indian arrows such as are used by the executioners in the West Indies – when a man is condemned to death they lay open his back with them and he dies of it – an instrument used by the Jews in circumcision, some very light wood from Africa, the robe of the King of Virginia . . .,

a passion of Christ carved very daintily on a plumstone, a large magnet stone, a S. Francis in wax under glass, as also a S. Jerome, the Pater Noster of Pope Gregory XV, pipes from the East and West Indies, a stone found in the West Indies in the water, whereon are graven Jesus, Mary and Joseph, a beautiful present from the Duke of Buckingham, which was of gold and diamonds affixed to a feather by which the four elements were signified, Isidor's MSS of *de natura hominis*, a scourge with which Charles V is said to have scourged himself, a hat band of snake bones . . .

At first we might think that there is not much difference between this and the lists drawn up by Cromwell's inspectors of the relics harboured by the religious houses of England, such as the coals on which St Lawrence was 'toasted', the paring of St Edmund's nails or the boots and penknife of St Thomas; but a moment's reflection will serve to bring out the differences. The Reformers' zeal reduces all the relics to inventory lists, but, as we have seen, it was not out of curiosity that the pilgrims went in search of the nails or the boots, but in order to enter the presence of the holy man. The coals, nails and boots stand for the saint himself, they evoke his presence. The pilgrim, in sixteenth-century England as much as in the countries of the Mediterranean in the late Roman Empire, was still 'going to a place to meet a person'; the visitor to John Tradescant's museum, on the other hand, was going to satisfy his curiosity and to be filled with wonder at the sheer quantity and variety of objects collected there.

What is so fascinating about Stirm's letter is the way no principle of organisation seems to operate, the way everything is jumbled together and exerts an equal pull on the viewer, a goose which grows on a tree in Scotland and pieces of the True Cross, the robe of the King of Virginia and a Passion carved on a stone. Stirm's only form of taxonomy seems to be the list. Just as the *Mona Lisa* in the Louvre is, for most visitors, merely 'Leonardo's masterpiece', to be wondered at before one passes on, so the elk's hoof, the mermaid's hand or the instrument used by the Jews for circumcision, wrenched from their original contexts, become objects of curiosity and nothing else.

This sense of the sheer abundance of objects in the world, no longer clearly linked to the single unifying story of Creation and Redemption, fuelled by the discovery of the Americas – and, later, of the Pacific and the African interior – is reflected in much of the writing of the period. The enormous lists of Rabelais and Ben Jonson, the overwhelming number of examples of melancholy in Burton's *Anatomy of Melancholy*, and of quincunxes and even 'vulgar errors' in Thomas Browne – these strike the reader as expressions of delighted exuberance at the sheer variety and richness of the world, but also as a kind of panic at the lack of any real or underlying order. And the attitude is not confined to the sixteenth and seventeenth centuries. Nicholas Thomas, in a fascinating article on Cook's Pacific voyages, describes, for example, a set of plates of curiosities in the British Museum, John and Andrew van Rymsdyck's *Museum Britannicum* (1788):

> The plates included Taylor-birds and wasps' nests; the Oculus Mundi, or eye of the world, a Chinese pebble that becomes transparent in water; a penknife with a gold tip, employed in an alchemist's sleight-of-hand; a brick from the Tower of Babel; 'A very curious *Coral*, modeled by Nature, in the form of a Hand or Glove'; Governor Pitt's brilliant diamond; and some weapons, including the Flagello, an unlawful instrument said to have been extensively used 'in the *Irish* massacre of King *Charles*'s time; though far be it from me to advance any thing that is not true'.

Thomas, like a number of recent scholars, is concerned to trace the uneasy cohabitation in such works of the scholarly and the prurient, and to alert us to the tension which, in eighteenth-century England at least, lay behind the use of the words 'curio', 'curious' and 'curiosity'. For Dr Johnson, for example, in his *Dictionary* (1755), to be curious is to be 'addicted to enquiry', but in his *Journey to the Western Isles* (1775) he reproves a writer who failed to ascertain the precise breadth of Loch Ness for being 'very incurious'. And this tension, Thomas shows with great erudition and sensitivity, is still there in the accounts (and no doubt the minds) of those first African explorers, John Barrow and Mungo Park who, at the turn of the nineteenth century, were still unclear whether their voyages were scientific, colonialist or driven simply by curiosity.

This was the time too, as we have seen, when the notion of genre first began to ring hollow, for genre depends ultimately on the sense that there is a place for everything and everything is in its place, that the world is ordered and art can reflect that order. It is no coincidence that the breakdown of genre, typified by Dr Johnson's criticism of *Lycidas*, should coincide with the rise of the novel, that genreless form which cannot make up its mind whether its ultimate appeal is to Truth or curiosity. Sterne plays with this (and with the reader) all the way through *Tristram Shandy*, notably when he orders the female reader back to the previous chapter and then stops to comment: 'I have imposed this penance upon the lady, neither out of wantonness nor cruelty, but from the best of motives . . . – 'Tis to rebuke a vicious taste, which has crept into thousands besides herself, – of reading straight forwards, more in quest of the adventures than of the deep erudition and knowledge which a book of this cast, if read over as it should be, would infallibly impart with them . . . – But here comes my fair Lady. Have you read over again the chapter, Madam, as I desired you?' (I.20).

In the wake of Krzystof Pomian's pioneering work on early Italian collectors and collections, many scholars have begun to explore the ways in which and the reasons for which national museums grew out of private and church collections, and to reveal the tensions that underlay them all between the claim to be advancing science, the pandering to curiosity and the growth of national pride. At first it was learned and wealthy individuals who collected. Humanist popes and Renaissance princes (often the same people) had started collecting classical statues in the fifteenth century. Others began to collect manuscripts and paintings. A number of Renaissance paintings (*Las Meninas* being among the last and greatest) actually depict the owner in the midst of his pictures, which are hung on and stacked against the walls around him. But by the nineteenth century the great private collections had been democratised. In Paris the Louvre, the ancient palace of the kings of France, became the Musée Napoléon; in Madrid the royal collections moved into the Prado, the first modern building to house an art museum. The Altes Museum was opened in Berlin in 1830, followed by Munich (1836), London (1838), Dresden (1855), Amsterdam (1885), Vienna (1891) and Moscow (1912).

But of course by the nineteenth century the mania for collecting had passed from the learned and wealthy, from emperors and governments to every middle-class child. This was the time when children's encyclopedias began to be published (the first, adult encyclopedias date from the previous century), and when fossils, butterflies, shells, pressed flowers and anything else one could think of began to be collected by children with the encouragement of their parents. Even in Egypt, on the fringes of the western world, in the middle years of the present century, I found myself caught up in the last stirrings of this mania, which was encouraged by the English books and magazines I read: *The Castle of Adventure, Swallows and Amazons, The Railway Children, Boy's Own Paper, The Eagle, World Sports*. I seem to have spent my childhood, when I was not playing football or tennis or taking part in swimming and athletic competitions, sticking pictures into scrapbooks, putting objects into boxes and labelling them or arranging them in specially prepared trays. My most treasured possessions were my collection of prehistoric flints, picked up in the course of walks and bicycle rides into the desert just beyond the confines of the little town where I grew up, and a scrapbook I made of the 1952 Olympic Games, in which every result of every heat of every event was lovingly included.

I'm not sure about stamp-collecting and train-spotting, but, by and large, it seems to me that the collecting mania was a richly educative one for the children who succumbed to it. One was, in a way, making something of one's own, and one learned in the process about archaeology and botany and history and geography and countless other subjects. Unfortunately, like so many other childhood passions, this mania for collecting can take on rather more sinister connotations when it passes undiminished into adulthood. One particular collecting mania indulged in by adults in the second half of the twentieth century demonstrates this only too clearly.

16   Possessing Power

To possess a relic was to possess power. As Peter Brown shows, at its origins the cult of relics was carefully controlled precisely so as to maintain the aura of the relic and to lead the pilgrim to recognise that distance was an essential component of *praesentia*. In this way the pilgrim would be led to place the relic within the context of a narrative at whose centre stood the death of the first martyr, Jesus Christ, the Son of God, who had died that mankind might be saved. By the time of the Reformation, as we can see from the many satires on relics and indulgences, on greedy clerics and gullible consumers, that sense of aura had all but vanished. A fragment of the true cross or a piece of saint's clothing could be thought as likely to protect you from death or disease as the magic girdle which Gawain so gratefully accepted. Yet the need to displace yourself and go to seek such relics still, to some extent, preserved the sense of *praesentia*, and the evidence at the shrines of the wonders and cures associated with the relics still conveyed powerfully the sense that, even if no one else did, God still cared for the poor, the excluded and the sick. A couple of centuries on and we see, in Tradescant's collection, and in the German visitor's response to it, that 'a small piece of wood from the cross of Christ' has become simply an object of natural curiosity, like a flying squirrel or the hand of a mermaid. What seems to happen in what French historians have called 'the history of everyday life' is that when certain social practices and assumptions are discarded as false and fantastic the needs they fulfilled remain and, with nothing now to acculturate them, become a source of pain and anxiety and the generators of dangerous dreams and desires.

Nowhere is this more in evidence than in the repulsive trade in Nazi memorabilia which has gone on unabated since the end of the Second World War and the collapse of the Third Reich. Robert Harris, who dug deep into this cesspit when researching for his book on the bizarre episode of the forged Hitler diaries, has come up with some startling figures (at least they startled me). 'It has been estimated', he writes, 'that there are 50,000 collectors of Nazi memorabilia throughout the world, of whom most are Americans, involved in a business which is said to have an annual turnover of $50 million.' In the United States a monthly newsletter, *Der Gauleiter*, keeps up to five thousand collectors and dealers informed of the latest trade shows and auctions. In Los Angeles one collector amuses himself in private by wearing Ribbentrop's overcoat. In Kansas City another serves drinks from Hitler's punch-bowl. In Chicago a family doctor has installed a concrete vault beneath his house, where he keeps a collection of Nazi weapons. In Arizona a used car salesman drives his family around in the Mercedes which Hitler gave to Eva Braun. In England, at his ancestral home, Longleat, the Marquess of Bath has gathered the largest collection of paintings by Hitler, worth, according to one estimate, ten million dollars. In the room with the paintings were to be seen as well 'a life-size wax model of Hitler wearing a black leather overcoat and a swastika armband . . ., Himmler's spectacles . . ., the Commandant of Belsen's tablecloth'.

In the course of his book Harris introduces us to a number of dealers and their collections. Konrad Kujau's 'included an almost complete set of Third Reich decorations, 150 helmets, 50 uniforms, 30 flags, and, according to Kujau, the largest collection of military jugs in West Germany'. In Goering's yacht, which he had bought, the journalist Gerd Heidemann set out Goering's dinner service, tea-cups, drinking goblets and ashtray. In the cupboard was Goering's uniform; the cushion covers were made from Goering's bathrobe. With the money paid to him for the Hitler diaries by *Stern* magazine Heidemann bought three hundred paintings, sketches and water-colours by Hitler, as well as Nazi party uniforms, banners, flags and postcards, as well as the actual revolver

Hitler used to shoot himself. Most of these, it turned out, were fakes, sold to him by Kujau, the forger of the Hitler diaries. But of course, as with Minoan and Egyptian antiquities earlier in the century, where there is a ready market fakes will abound.

Before he shot himself Hitler tried to ensure that his personal papers would all be destroyed. But already those around him had started collecting. When, on 24 August 1945, American agents raided a house in Schladming, Austria, they found Eva Braun's private photograph album, the notes she had made of her letters to Hitler, and a Hitler uniform. In October they carried out another raid and found twenty-eight reels of colour film, Eva Braun's home movies of her life with Hitler.

As one can see from all this, it is difficult to tell where devotion to Hitler and sheer greed begin and end. As with many late medieval relics, there is a grey area here, and probably a few of the dealers and collectors could themselves say truthfully whether they were in it for the money or for the sense that they were handling what was, after all, a part of history. There is also the sense, exploited in a specific genre of post-war art, film in particular, of the thrill of transgression, of that stagey fetishism which Proust analysed so lucidly and which has rightly been described as camp or kitsch, though unfortunately of a rather more public and dangerous kind than that which the young Marcel so surprisingly came upon that day in Monjouvain.

Konrad Kujau managed to carry on his lucrative trade in forged Nazi memorabilia for many years, until the grandiose scheme of the Hitler diaries caused him to overreach himself and he was finally caught. Harris concludes his book with a question: 'Why should anyone pay $3500 for a few strands of human hair of dubious authenticity?' And he answers:

Because, presumably, *he* might have touched them, as he might have touched the odd scrap of paper, or painting, or piece of uniform – talismans which have been handed down and sold and hoarded, to be brought out and touched occasionally, as if the essence of the man somehow lived on in them. The Hitler diaries, shabby forgeries . . . were no different. 'It was a very special thing to hold such a thing in your hand', said Manfred Fisher, trying to explain the fascination which he and his colleagues felt when the first volume arrived. 'To think that this diary was written by *him* – and now I have it in my grasp . . .'

But this, after all, Harris adds, was 'a phenomenon which Chaucer's Pardoner, six centuries ago, with his pillow cases and pig's bones, would have recognised at once'.

I think though that we should be wary of making too direct and immediate an identification of Nazi relics with those of the late Middle Ages (just as we should of calling the exhibits in a modern museum 'secular relics', as some historians of museums have done). In Chaucer's tale we are still in a world which, for all the abuses it could see in the Church and its ways, still believed that Christianity was somehow coterminous with the universe. The Pardoner is a tragic figure as Kujau is not, partly because what he has to offer is after all a promise of healing and redemption, and partly because his insistence on the fact that his only creed is cupidity is always shadowed by his refrain: *radix malorum est cupiditas* – the root of all evil is cupidity. In other words, he knows that he himself is damned, as Kujau never does.

The power of the relics Kujau handles depends partly on the horror associated with them and partly on the sense that by handling them one is taking part in a giant play, a play which for a while threatened to overwhelm reality, a play in which the very impulse towards annihilation of others and oneself was for a while given free rein, but which has long since been shown up for the hollow sham it always was. Or – and this is the ultimate source of the *frisson* which handling such things no doubt conveys – was it?

17   The Jewish Bride

They look at us but they are thinking of each other. Despite the occasion and the finery of their dress, they – she in particular – have that faintly melancholy expression one sees in so many Jewish faces. He sits on the wooden arm of an elegant divan and draws her towards him, his right arm round her waist and his head inclined against her shoulder, so that his cheek seems just to touch the gloved hand she has put up almost protectively. In his left hand he holds his grey top hat and his left foot is thrust forward towards us. She, standing there beside him, does not seem to succumb entirely, partly because he is so precariously perched but also for deeper, more complex reasons, which she would probably not be able to put into words. Her white dress flows down to her feet where it forms a pool of white that covers his right foot, and in her right hand she holds, rather nonchalantly, a bouquet of flowers, their heads pointing towards the floor, which, with the end of the long girdle that runs parallel to it down the left side of her body, introduces an element of fantasy and naturalness into this otherwise rather formal wedding photograph.

They are my grandparents, and the photo dates from the turn of the century. It must resemble thousands if not millions of other wedding photographs of that date. Yet even in its slight stiffness what it registers is above all the trust between these two people as they touch yet remain slightly apart from each other. Though the stance is a little different, it is not all that far removed from Rembrandt's great painting of *The Jewish Bride*, which similarly conveys a moving combination of tenderness and formality – tenderness in spite of and as a result of the formality. In Rembrandt's painting the man's hand lies protectively just beneath the woman's bosom and her hand rests lightly on his, only four fingers making contact, the merest touch. In the photo, on the other hand, his arm goes protectively round her waist, his hand appearing just above her elegantly curved hip, and she leans towards him, but the message is the same: these two people have entrusted themselves to each other and, for the moment at any rate, they are slightly overawed by the enormity of what they have done; for one long moment touch and distance are combined.

And of course it is a long moment. Because of the changes in the techniques of photography an image such as this, which is in many ways closer to a painting than to a modern snapshot, is simply no longer a possibility. In our much less formal world, too, wedding photographs tend to put one in mind more of a play or a charade than of real life: even when the suits and dresses are not hired for the occasion they tend to be worn awkwardly by people more used to dressing in jeans and pullovers and there is always the sense that this is a comedy put on for other people rather than a moment of significance for themselves. But here there is no play. There is a continuity, we feel, with what has gone before and what is to come, even if the moment itself is of supreme importance. These clothes belong to them; the elegant divan and the heavy drapes of the curtains which form the background do not belong to a photographer's studio but to the drawing room of her father. Their expressions acknowledge and take responsibility for what is laid open to the camera: they are what we see and what they have accepted to reveal. In a secular age which nevertheless still lives to the rhythms of sacramental tradition the photo here could almost be said to have taken over the function of the church or synagogue ceremony: two people here make public their lifelong commitment to each other.

Andrew Graham-Dixon has recently written of *The Jewish Bride*:

It is one of the great paintings of love, of unaffected tenderness, in Western art. It is not idealised or dramatised. Rembrandt's figures do

99

2. *(right)* Alexei Rabinovitch
and Nelly Rossi, Cairo 1907.

3. *(facing page)* Rembrandt,
The Jewish Bride, c. 1666.
(Rijksmuseum, Amsterdam)

not adopt the stock theatrical postures of the amorous in art. Love,
here, expresses itself through the subtlest language of face and body;
through expressions that seem charged with emotion and, also, a peculiar sense of momentousness, as if the issue of this union might be as
much a cause for apprehension as celebration; through the inclination
of the man's body, its slight but immensely affectionate lean from
the vertical; through the contact of his hand with the woman's, and
the lightness of that contact, so skilfully (and inexplicably) expressed
in paint.

That seems exactly right, and if the photo is inevitably less charged, if
the faces of the bride and groom seem perhaps even slightly bland,

slightly lacking in expressiveness, that is only the difference between photography and paint; not just the difference between the anonymous photographer and Rembrandt but also our different reactions to being photographed and painted, the way we freeze a little to try instinctively to protect ourselves from a photographer, whose act, however well we know him, is always felt as predatory. But in both images touch is at once absolutely natural and, as Graham-Dixon says, momentous. It establishes distance as much as closeness. It shows the relation between the public and the private spheres, between irreducible singularity and the commitment to another. For both couples are anything but starry-eyed. Both seem to see clearly the vicissitudes that lie ahead. But touch is, in both instances, the index of their confidence and trust. Touch establishes the boundaries of each and their dependence on each other, neither perpetual solitude nor perpetual merging but an acceptance of difference in their free decision to make their lives together.

18   First Steps

It was just at the time when newly married couples throughout the western world were having their photographs taken in postures and clothing similar to those of my grandparents that we entered the period that has aptly been called the Age of Suspicion. For the one thing that seems to unite all the greatest artists and thinkers of the later nineteenth century – Marx, Kierkegaard, Dostoevsky, Flaubert, Nietzsche – and of the twentieth – Proust, Wittgenstein, Kafka, Freud, Eliot, Picasso, Schoenberg – was their suspicion of the notions they had inherited from the Enlightenment and Romanticism – Progress, Reason, Imagination, Art, Glory, Human Nature, etc. – and their desire to strip away false façades and lay bare the needs and motives that had given rise to them.

Yet, among the artists at least, this desire went hand in hand with a profound sense that the ethos of suspicion was always in danger of sweeping away the precious and genuine along with the false and the shoddy, with an awareness, largely instinctive, that if our relation to the world is to be one of unwavering suspicion then we will have wounded ourselves more deeply than we realise. Van Gogh, quite as iconoclastic as Nietzsche in his way, could nevertheless write to his brother about 'what only Rembrandt has among painters, that tenderness in the gaze . . . that heart-broken tenderness, that glimpse of a superhuman infinite that there seems so natural'. Unmasking the sentimental, the hypocritical, the oppressive, the shoddy, can, indeed must, co-exist with a recognition of the true and the genuine.

Picasso, less romantic than van Gogh but equally responsive to those aspects of our lives that call for celebration as well as those which need to be unmasked, has left us a whole series of icons of trust, particularly that which exists between mother and child, at the same time as he wittily undermines the pretensions and bad faith that so often underlie nineteenth-century depictions of this theme. Two pieces of sculpture in particular come to mind. One of them is so famous that it is now difficult for us to read it. In Vallauris in 1952 the seventy-year-old artist found a toy car which, in a flash of inspiration, he saw as the broad-nosed face of a monkey. The finished sculpture, however, provides us with more than the simple play of wit embodied in that perception. For it is an image not of a car/monkey but of a mother and child. The tiny baby monkey, clinging to the mother's massive chest, completely alters the meaning of the car/face above it. Quite as much as any madonna and child, what the sculpture suggests is the child's need for dependence and the mother's tenderness, clumsiness and pride. Had this been a work of Duchamp's none of this would have emerged. Duchamp would have made the connection between car and monkey-face and been satisfied to embody that flash of wit. Indeed, he would probably have felt that anything more would destroy the purity of the initial conception. For Picasso, on the other hand, the flash of wit – this car, seen from above, is the face of a monkey – is only the beginning.

Two years earlier he had welded together another piece of sculpture made from found objects. This too is a mother and child, but the tall, stately mother is pushing a pram this time, in which sits a tiny infant. The mother's hands just touch the enormously elongated handle of the pram, her head on its long neck looks up and out, but again the piece is much more than a playful *tour de force*: a spark flies between the child in the pram and the mother standing proudly above it, a spark which elicits sympathy and understanding from us, however 'unrealistic' the mother and child may be. As always with Picasso the technical skill and witty insight are not there for their own sakes or even to make a general point about perception or art, but for the sake of the human content.

4. (*right*) Pablo Picasso, *Mother and Child*,
1952. (Musée Picasso, Paris. Photo © RMN.
© Succession Picasso/DACS, 1996)

5. (*below*) Pablo Picasso, *Mother with
Pushchair*, 1950. (Musée Picasso, Paris.
Photo © RMN. © Succession Picasso/
DACS, 1996)

6. (*left*) Pablo Picasso, *First Steps*, 1943. (Yale University Art Gallery, New Haven; gift of Stephen C. Clark B.A. 1903. © Succession Picasso/DACS 1996)

7. (*below*) Rembrandt, *Two Women Teaching a Child to Walk*, *c.* 1635–7. (British Museum)

There is nothing in the least witty, though, about the 1943 painting, *First Steps*, in which a grossly distorted mother, bending over an equally distorted infant and holding both his hands, is helping him to learn to walk. Hockney once remarked, *apropos* of this painting, that even great artists rarely add to the repertoire of motifs that have dominated the western painting tradition, but that here Picasso does just that. The theme of the child's first steps is after all a universal one, since the moment the child discovers he can walk on his own is one of the key moments in every human life – yet how often, asks Hockney, has it been portrayed in art? (I can't believe he did not know that among the drawings by Rembrandt owned by the British Museum is one, dated 1635–7, which shows two women teaching a funny upright little child to walk, and one, dating from the 1650s, which shows a mother similarly engaged with her child. Yet his point of course still holds: the iconography of first steps is an extremely rare one in art.)

Here, in the Picasso, the post-Cubist distortions convey, in a manner figurative realism never could, the mother's love and concern, her hopes and fears, and the child's equally powerful mixture of terror and excitement. She thinks of nothing but him and his well-being, he thinks of nothing but his body and the freedom that beckons. She is about to let go the two hands she holds in hers, while the absurdly enlarged feet of the child are already on their way to freedom, the left raised so that we see the toes and most of the sole from below. A miracle no less great than that of the Resurrection is about to occur, the miracle that takes place in every human life when we are forced to let go and find that we can actually make it on our own, when we reconcile ourselves to letting go and find that those we have nurtured can actually make it on their own. Trusting that her hands will hold him should he stumble, the child discovers that he can retain his balance, that his legs will hold him upright and that he is no longer attached to his mother; trusting that his feet and legs will carry him, the mother lets him go with a mixture of pride – look at what he can do! – and sorrow – this is an index of things to come.

That charged moment enters into dialogue with Proust's equally charged depiction of the moment when the child discovers that he is not the centre of the world, that even his mother has a life and interests of her own, and in that instant discovers his profound dependence on her. In the Picasso the instinct to cling is overcome by the need to set out; in the Proust the need to know that the parent can be relied on in every circumstance turns the steadying touch into the anxious grasp, with inevitably tragic results. We have all of us experienced both emotions, and they have marked us for life. Who is to say which is the deeper or which has marked us more?

19 Kinetic Melodies

One of the ways in which the masters of suspicion worked was to reveal to us that what we had taken to be natural, a 'given', was in fact man-made, the result of choices and decisions made by individuals and institutions. Thus Marx unmasked the workings of capital, Nietzsche the workings of morality and Freud the workings of sexuality. Where the Enlightenment had seen all men as essentially unchanging and human nature as universal, the nineteenth-century masters of suspicion set about exploring the genealogies of morals and social institutions with the aim of freeing men from forms of bondage to which they did not even know they were subject.

The work of Oliver Sacks has stood at a curious angle to all this. He too is an unmasker, he too is concerned that we should see that much of what we take for granted needs to be examined critically. But his aim seems to be rather to make us aware of the extraordinary nature of what is unmasked, to fill us with wonder again at the almost miraculous nature of so much in our lives that we tend to take for granted. In one brilliant and illuminating article and book after another he has helped us to understand the complexity of the way the human mind and body function and of such apparently simple things as standing, walking and talking.

He has done this negatively, by exploring what happens when those things we normally take for granted no longer seem to function, when, for one reason or another, we cease to be able to cross a room, ask for a glass of water, remember what our husband or wife looks like. As he describes, with a novelist's empathy, patients who seem driven either to rush frantically forward or to stop dead in their tracks, either to gabble incoherently or to search hopelessly for the most ordinary words, we come to realise how amazing it is that we can walk up stairs without even thinking about it, communicate with our friends, understand what we are told and act upon it.

This was not of course his primary aim, which, in the time-honoured way of medical papers, was designed to explain illness where he found it and to suggest cures. Nor of course was he the first to bring empathy, imagination and a sharp eye for detail to bear on neurological problems affecting human beings. His own mentor, the great Russian neurosurgeon A. R. Luria, and such late nineteenth- and early twentieth-century medical experts as Sir Charles Bell, the author of a famous book on the mechanism of the hand, and Henry Head, the influential author of *Studies in Neurology* (1920), had already shown the way. But Sacks' unique *writerly* gifts brought this branch of medicine into general public consciousness, first with his extraordinary study of the aftermath of the great influenza epidemic of 1919, *Awakenings*, and then in a host of case studies published in non-medical journals, and later collected into books, such as *The Man Who Mistook his Wife for a Hat* and *An Anthropologist on Mars*.

But it was with a book he published in 1984 about an experience of his own, *A Leg to Stand On*, that Sacks brought to the centre of our attention the notions of proprioception and kinetic melody (a term coined by Luria). Having broken his leg on a Norwegian mountain during a holiday, Sacks discovered that, due not to a neuropathy but to a nerve and muscle injury, he no longer recognised the leg as his own. Not only could he not feel it, he had forgotten what it was like to have a leg. He recalled then a former patient of his who, in his half-sleep, would sense an alien presence in his bed and, in a fit of panic, struggle with it until he would eventually manage to hurl it out of bed – only to discover that it was his own leg and he was now lying on the floor. This had been totally baffling to the doctors and nurses at the time, but as Sacks struggled to understand what was happening to himself he began to ponder again the notion of body-image or proprioception, which has long been known in medicine but improperly understood.

Proprioception is our totally intuitive sense of our own bodies, without which we could not function at all, a kind of sixth sense which keeps our other senses in a balanced relation to each other, rather like the pentangle on Gawain's shield. More basic than sex or even desire, proprioception is the body's own sense of itself as occupying space and as active in that space. Deprived of this sense we are not merely helpless, we cannot even stand – let alone move. And yet so deeply is that sense buried within us, so much is it the ground of our being, that we take it totally for granted and only become aware of its crucial function when something goes wrong with it. Kinetic melody is the complement of this, our instinctive ability to write, sing, dance and so on, activities we could never do if we had to think through every movement of the hand or foot.

Sacks eventually recovered the use of his leg and the sense of his leg as his own, although it was to be ten years before he was able to write in any coherent way about his totally disorientating experience. Others have not been so lucky. In 1991 a pupil of Sacks, Jonathan Cole, published a book which told the story of a man who, at the age of nineteen, in the aftermath of a bout of gastric flu, was deprived of all sense of his body from the neck down, and never recovered. *Pride and a Daily Marathon* is Cole's extraordinary story of Ian Waterman and of his fight to lead a normal life despite the fact that he would never regain his proprioception. The book follows Waterman as he quite literally wills himself to function normally, which means, in effect, to imitate every detail of the movements of those he sees around him, since without a sense of his own body nothing can come naturally to him. At every second, therefore, he needs to be totally vigilant, lest, for example, a slight incline in the path, which we would adjust to without even becoming conscious of it, catch him unawares and cause him to stumble and fall, or the hand he moves towards his cup of tea fly suddenly sideways and strike his neighbour. Since when he cannot see his body he has no sense that it exists, darkness is the great enemy. Once, when he was alone in a lift, the lights failed and he was found in a crumpled heap on the floor when the doors were finally opened.

Ian Waterman's appalling plight, allied to his remarkable will and ability to express himself, along with Jonathan Cole's patience and empathy, makes one realise the vast range of complex things we do all the time without our even noticing. When we speak to others, for example, we speak with our bodies as much as with words. 'Since his illness', writes Cole,

Ian has lost all [his] unconscious body language and is no longer able to employ the myriad body postures and movements we use in communication and deception . . . Ian is aware of this lack and, as always, has evolved ways of circumventing it: 'Sure, I realise that, say, to incline towards someone when sitting is a sign of affection. Remember, I was all right until the age of 19, so I had learnt all this movement behaviour by then. Now I sometimes use it, but only when sitting down, which allows me to do it without fear of falling, and I always have to think about it. I have to decide consciously to use my hands to move with and emphasise my speech.'

Cole comments on this:

He has remembered and relearnt a limited repertoire of non-verbal communication which he now uses consciously when appropriate. To sit and talk to him is to see the arms and hands lifted up and down in front of him elaborating and extending the points made. On closer inspection, however, we realise that though the points are emphasised with the hands the fingers remain relatively still.

I cannot help thinking about Kafka when I read about Ian Waterman. Not only the ape in 'A Lecture to an Academy', who deliberately and self-consciously learns the gestures and speech of men so as to escape the fate which would otherwise await him, of living out his life in a cage, but also Kafka's own remark in a letter to Milena: 'Nothing is granted to me, everything has to be earned, not only the present and the future, but the past too – something after all which perhaps every human being has inherited, this too must be earned, it is perhaps the hardest work.'

Of course in one sense there is nothing at all in common between Ian Waterman and Franz Kafka, and it may seem almost insulting to the former even to think of comparing them. Kafka's was a set of feelings, Waterman's is a physical condition, and until the onset of his tuberculosis Kafka was a perfectly ordinary human being who, we might think, with a little bit of luck might have led a happy enough life. We might

111

even feel that his continual moaning is an indication only of self-pity and deserves nothing but contempt. But I don't think our fear of sentimentality and our respect for the horror of Ian Waterman's situation and his nobility of spirit in fighting it with such determination should inhibit us from considering the analogy. For what such a consideration reveals is that tradition, from which Kafka feels himself so totally and irrevocably cut off, functions in a way that is precisely analogous to proprioception – it is an unconscious set of habits and practices which allows us to function and which only reveals itself as the vital thing it is when we suddenly find ourselves deprived of it. Robbed of proprioception by his illness, Ian Waterman finds that he can function only by imitating, without feeling it, what others do instinctively; robbed of tradition by complex social changes and by his personal situation, Kafka finds that each step he takes, in both life and art, has to be a conscious one, each gesture and phrase to be planned and chosen with care and deliberation. In both cases there is the profound desire to go on, not to succumb to this thing that has overtaken them, and in both what keeps them going is pride and a kind of trust against the odds. For both of them every day is a marathon, a test of endurance and resolve.

Ian Waterman's story is of course neither allegory nor exemplum. It is itself. He is himself. Nevertheless, it helps us understand the nature of the artistic difficulties encountered by an Eliot, a Kafka, a Stravinsky, a Picasso, as they tried to live their lives as artists in our century.

At the same time the study of proprioception, of that kinetic melody or the body's ability to find its own rhythm, its knowledge which is beyond any knowledge which conscious thought can offer, should also help us understand something the neurologists, for obvious reasons, are not particularly interested in, the difference between being heavy-handed and having a light touch, between being an indifferent pianist or footballer or tennis player and a great one. For while most of us are indeed blessed with the ability to walk and run and strike the right notes on a piano and hit a ball, some are more gifted in these spheres than others, and even those not specially gifted have experienced, at one time or another, moments when they have felt themselves to be running, swimming, hitting a ball or striking the keys of a piano as, so to speak, we were all meant to run, swim and strike a ball or the keys of a piano. At such moments we feel ourselves to be *more ourselves* than is normally the case, more actively a part of the world. At such moments, one could say, we understand instinctively what it means to exist, that miracle which most of the time we simply take for granted.

20   Kinetic Melodies (2)

Football, in Egypt (where I learnt to play it at the age of five), as in South America, is a game of touch and skill. The reasons are obvious: the ground is hard and dry, the ball is always light, and such conditions favour the quick and nimble rather than the bulky and powerful, the good dribbler prepared to take people on rather than the heavy tackler intent only on blocking the progress of the opposition. Those were the days of 2-3-5 formations, 2 backs, 3 halves, 2 wings, 2 inside-forwards and a centre-forward. No one had heard of total football, neat triangles, Christmas trees or diamonds: the idea was to get the ball to the forwards as quickly as possible and let them get on with it. Watching us, present-day managers would have had a fit: when the opposition was attacking we forwards hung around the half-way line, waiting and hoping. Once we had the ball it was up to us to make for the goal as quickly as possible and score. They were high-scoring games, I remember, with twenty-two small boys on a full-sized pitch and a diminutive goalkeeper defending a full-sized goal. And they were never boring.

There were magical days when I could do no wrong, scoring three or four of the six or seven goals which earned us a narrow victory and setting up at least two of the others. On days like that it seemed so easy to swerve round the half-backs, cut inside the backs and send the ball high into the corner of the net. There must have been bad days, of course, when nothing would go right, when I never got the ball or could do nothing with it when I did get it. But my memory has blotted these out as effectively as mothers are said to blot out the pains of childbirth.

What does having 'a good touch' mean on the football field? It means having an instinctive sense of the ball as it comes to you, its speed and trajectory; it means knowing instinctively how to position yourself to receive a pass most effectively, when to linger and when to accelerate; it means feeling the ball as a part of yourself not just when it's at your feet but when it's at the other end of the field; it means internalising the pitch, sensing the game as good chess players are said to do the board before them, as a series not of static positions but of lines of force.

No one knows how they are going to perform when they step on to a football field or a tennis court. That is part of the beauty of sport, why it is life in microcosm, life compressed and heightened by rules and by the limits of space and time. Once you are out there you are on your own, the coaching, the training, the team-talks, the personal advice no longer relevant; only in the course of the game do you discover how good or bad you are (on that particular day).

This is even truer of tennis than it is of football, for in tennis you can't complain that you didn't get the right passes or lost because someone else was not on form or made a crucial blunder. In tennis there is no one but yourself to blame, though it does happen that one comes up against an opponent who, for the duration of the match, seems himself to be caught up in a dream of perfection and can do no wrong.

We talk of touch players in tennis much more than in football and all those who follow the game have a list of such players. Mine would include Okker, Goolagong, Nastase and McEnroe. But all the great players have touch. They all internalise the court instinctively, so that the game is not so much played out there where the spectators see it as, somehow, 'in there', in their bodies. Watching Hoad or Laver at their rampant best was like being invited into a dream. They seemed always to pick the right shot to play and it seemed that they not so much ran for the ball as that the ball was inevitably drawn into the centre of their racket.

The beauty and pain of both playing and watching tennis, though, has a great deal to do with the scoring system, which makes it very difficult for anyone to keep this up for a whole match. Hoad managed it when he demolished Ashley Cooper in the 1957 Wimbledon final (the first game of tennis I ever saw on television), and Connors managed it when he thrashed Rosewall in the 1973 final. But most of the time no one can take complete control, as one would in a race or a football match. There is too much time between points and between games. However hard you try you cannot keep yourself from thinking, from wondering if your touch is going to desert you or your opponent is finally going to find his. So many dreams turn into nightmares, tests not so much of skill and stamina, as in other sports, but of your whole balance and co-ordination. You start to become conscious of your touch and that is probably the moment when it begins to desert you. Whether you have begun to grow aware of it because it has already started to go or whether it starts to go precisely because you become aware of it no one can tell. But it is a curious and well-known fact about tennis that to win the first set 6–0 may not be the best way of ensuring an eventual victory. I suspect more matches are won from 0–6 down than from 3 or 4–6.

Tennis is a game where real time plays a key role. In football one may rue a missed opportunity but there is little time to think about it. In tennis the half-chance not taken, the break-point muffed, goes on haunting one and can easily lead to that dread tightening of the arm, that sense of unease which no amount of will or exertion can dissipate. That was Borg's great strength, apart from his extraordinary balance and speed of foot: a point played seemed to be over and forgotten as soon as it was done, leaving him totally free to concentrate absolutely on the next one.

A game which had looked comfortably won one moment can seem irretrievable only ten minutes later, and one wonders miserably, between points, what on earth one can do to get back the touch which was so effortlessly there just minutes before. So Adam must have wondered about the eating of the apple as he sweated away miserably, trying to scratch a living from the soil, and so Marcel must have felt on waking up to recall how, the evening before, he had managed to inveigle his mother into spending much of the night reading to him from his favourite book.

A few years ago I took up Aikido. I had twisted my knee playing six-a-side football long after I should have given up the game, and it wouldn't stop hurting. Every time I tried to play tennis the injury flared up again. The doctors prodded and X-rayed but couldn't find anything wrong. Finally a specialist suggested that the only solution would be to go in to hospital for an exploratory probe. The idea did not appeal to me, and when I mentioned it to a former pupil who had spent a year in Japan and, I knew, practised some sort of Japanese martial art, he said Aikido would be just the thing for the knee. A lot of it was actually done on the knees, he said, and I would thus gradually build up the muscles in that region. I had never much liked the idea of bowing to portraits of the Master and throwing opponents over my shoulder (or, for that matter, being thrown over someone else's shoulder), but he merely smiled when I said all this and suggested I come along to the Dojo one day and have a look. So as not to appear churlish I went, and was pleasantly surprised. Although classed as a martial art, Aikido is non-aggressive. You work with a partner, not an opponent, and, though there was quite a lot of bowing, there seemed to be much less throwing over the shoulders than I had imagined. I decided to give it a try.

What I learned from my few years of Aikido has helped me to understand many things, for what it teaches applies to most activities, both physical and mental. The one key lesson is that the ideal state is one where you are both utterly relaxed and utterly concentrated. This is a notion which is alien to most of us in the West, who tend to think of relaxation and concentration as opposed states. Yet we have all of us experienced moments when this was, in fact, not so, when the two did co-exist in a fruitful balance. Certainly this was what happened when I was running and swimming well and what now happens when I am writing well. But in our culture concentration implies stiffness, tension, all sorts of energies repressed or kept at bay, while relaxation means just the opposite: slackness, letting go, doing nothing. Virgil and Milton are the great poets of this opposition. For them to relax is to give in to temptation and what is important is to be constantly vigilant. Milton's heroine in *Comus* prefers to turn into stone rather than succumb to the seduction of the nature spirit, and Milton's very style suggests a huge effort of the will, an imposition of order on a dangerously unruly language. The power of his writing makes this opposition almost believable, but my own experiences of swimming, of Aikido and of writing convince me that the antithesis he develops is a false one. There is nothing esoteric about the famed nonchalance of the Zen archer or what our Aikido teacher tried to instil into us, but, like all such things, it is a simple lesson which is extremely hard to learn.

In Egypt, when I was a child, we used to train every afternoon from May to September in the open-air pool next to the football field. Dotted through the season were a number of competitions, divided into age groups from under twelve to under sixteen. At times training seemed wearisome and repetitious. At times it felt as though one was going backwards in terms of both technique and speed. But if one had put in the right kind of training and the right amount of it, and if the coach had got his timing right and one peaked for the most important competitions, then the actual races were wonderful occasions, even though one invariably felt terrible beforehand.

The key here too was to stay relaxed while concentrating totally. If you tightened up you were done for, and yet a lapse of concentration at the start or the turn or in the final few strokes would ruin the work of months of hard training.

In swimming one is using one's whole body: the arms pull and it is vital to keep the strokes as long and relaxed as possible; the legs beat, the kick starting in the region of the abdomen and going down through the thighs and calves to the ankles in a whiplash action; and breathing has to stay as measured as possible. The faster one goes the higher up in the water the body rises and the easier it is on the arms, back and chest and so, of course, on the breathing. Yet the faster one goes the more energy one is using up and thus the harder it is to breathe and the greater the pain felt by the body.

When everything clicks, as it did for me in one memorable race, you no longer feel yourself as made up of arms, legs, torso, neck and head. Instead you are a single living entity, a centre of energy without any definable outline. You are moving fast and your heart is pounding, not so much with the effort as with the excitement; in spite of this you feel the power you are generating as you churn through the water and feel too that you are staying beautifully relaxed. For this timeless moment body and mind are one in the effort being made and the satisfaction achieved.

It was the final of a national one-hundred-metres free-style race. My problem was that I always tended to start off a little too slowly, afraid that I would not have anything left in the last twenty metres. But I had registered only the third fastest time in the heats and I knew that if I was to give myself a chance of winning I would have to take a risk with the finish and start off as though it were a fifty-metre sprint. At the turn (the championships were held in an Olympic-sized pool in the open air) I could sense that I was fractionally ahead of my two rivals, and wondered if I had set off too fast this time and would now start to pay the penalty.

But I took the turn well and kicked down the second length as I had done so many times in training, knowing that it was now just a matter of keeping my form till the finish. I could see the dim shape of my nearest rival alongside me in the water and sensed that he was making up ground. But at seventy metres I knew I was still marginally ahead. This was when suddenly everything started to hurt. But this was also where the hours and hours of training started to pay off. I had been in this sort of pain before and knew I could go on without tightening up, keeping the leg-beat deep and the breathing regular. I could see the crowds on the side of the pool every time I turned my head to breathe and could sense, rather than hear, the shouts. I knew my rivals hadn't actually got past me, but whether I was still ahead of them or not was impossible to determine. Then the dark mass of the end of the pool loomed ahead and it was just a question of digging down into the last reserves, not taking breath for those final few strokes even though my lungs felt as though they were about to explode, and hoping that I would touch the wall at the end of my stroke and not have to glide in or take an extra short stroke.

Afterwards it is several minutes before anyone knows the result. All one wants to do is breathe, calm the terrible pounding of the heart, the heaving of the chest. Then one is aware that the time-keepers and judges are in deep conversation. At least, I thought, it was close, at least I gave them a better race than the last time. With luck I may have improved my own time for the distance and one can't ask for more than that. Except that one can and does. However good the time, there is no substitute for winning.

But why should that be? Surely, one might think, swimming well, swimming to the best of one's ability, feeling the body alive, at ease with itself, is reward enough. At such times, after all, one is at the opposite pole from the prisoner in solitary confinement, from the ambiguous pleasures of addiction and perversity. At such times it seems that if the body has any destiny it is this. And yet, however much I might comfort myself by saying that I had, after all, set a personal best time, that I had simply been beaten by the better man, to have come so close and not won would have been a bitter disappointment. Of course to win when one has not swum particularly well, to win because one's opponents were themselves not on form, is not particularly satisfying either. But there is no doubt that the months of training and sacrifice only seem to pay off fully if one wins. Is it because to be beaten suggests that one is psychologically not up to the challenge, even though one may be physically well prepared? I don't know. All I know is that swimming or running faster than one has ever done before *and* winning is the ultimate satisfaction for any athlete. Then it becomes a pleasure to look back over the race and remember how it unfolded, how one felt at every moment, and what, in retrospect, were the turning-points. Then one recalls the bare minute it took for the race to unfold as a minute when one had somehow spoken and said all one had ever wanted to say. The pleasure that brings with it stays with one for ever, even though at the time it passes quickly and one returns to preparing for the next race.

21 Walker and World

To walk in intense heat with not a breath of air stirring
requires a steeling of the will. I recall days in Egypt on
which, after nine o'clock in the morning, it required real
determination simply to step out of the house and into the
sun. To walk in the pelting rain or a violent wind, as one
so often has to do in England, though, is not much fun
either. One can feel invigorated and perhaps virtuous
when the walk is over, but the walk itself is something to
be got through rather than enjoyed. I don't know how
often I have had to walk my dogs over the downs in that
sort of weather, cursing England, the dogs, and myself for
ever having them. Even the dogs look miserable and there
is probably only one thought in all our minds: how quickly
can we get back home again?

But when, in England, the air is crisp and there is only a
mild breeze blowing, when there is springy turf under
foot, as there is on the South Downs and in the numerous
beautiful limestone regions of the British Isles in the
summer months – then one ceases after an hour or two to
think about anything in particular, one becomes an
embodiment of kinetic melody. Unlike swimming, where
the body is fully co-ordinated only when one is moving at
speed, in walking the body finds its true rhythm only when
there is all the time in the world. Normally we either find
ourselves lacking the energy to carry out some daily task,

or having more energy than we quite know how to use. But walking for some distance on a good day in England makes all such frustrations vanish as though they had never been.

Think of what it would be like to walk if one's feet did not touch the earth, if the breeze did not touch one's face and arms. Of course nowadays we no longer have to imagine it. We can watch films of astronauts in space capsules or on the moon, or read their comments about what it feels like to move in an atmosphere devoid of gravity. It seems to be an eerie and not particularly pleasant experience.

Men, of course, have always dreamed of flight. But why should we want wings when we have feet?

The hard climbs in the Alps, three or four hundred metres, sometimes as much as a thousand, straight up and without respite, are not exactly fun. But they form part of a whole, and better than the view from the top is the feeling in one's body as one finally reaches the top and starts to walk along a ridge. It is as though, for a walk to be fully satisfactory, there have to be hard bits as well as easy bits, climbs as well as descents. Not that descents are particularly easy. I remember on my first visit to the Dolomites taking a ski-lift to the top of the highest mountain in that region and then walking back down to the hotel where we were staying. It was a nightmare. Only someone as ignorant of the mountains as I was then would ever have planned such a walk. My toes and ankles were in a terrible state by the time I got back and I didn't even have the satisfaction of achievement.

The pleasure of a walk does not lie in having mastered something or pushed oneself beyond what one thought one could do. No two walks are the same, and it is the feeling that this unique event is happening now, that I am part of it, plus the simple feeling of well-being generated by a good walk which is the important thing. Instead of feeling 'more!' or 'enough!', instead of feeling that the end is too impossibly far away or too ridiculously near, there is only the sense that, for the duration of the walk, desire and its satisfaction are one.

A few years ago in the Alps I acquired a stick. A modest piece of wood, curved at one end, with the knobs left by the removal of the twigs barely smoothed over, the whole varnished a dark brown and fitted with a metal point. I am not sure how much help it has been in the mountains, but it has become almost as indispensable to me as a dog. Now, even on the gentle South Downs, with nothing but soft chalk underfoot, I find myself missing something if, having set out, I discover that I have left my stick behind.

There is a way of deploying a stick on a walk, down, twirl, one two three four, down, twirl, one two three four, which comes as naturally as walking itself. The stick becomes an extra limb and the twirl confirms one in the act of walking, breathing the air, responding to the earth underfoot.

I have often thought, as I walked with my stick, down, twirl, one two three four, down, twirl, one two three four, of that page in *Tristram Shandy*, which cannot be paraphrased, only reproduced:

Whilst a man is free – cried the Corporal, giving a flourish with his stick thus –

A thousand of my father's most subtle syllogisms could not have said more for celibacy. (IX.4)

The question at issue is whether Uncle Toby's desire and need for the Widow Wadman is greater than his fear of being trapped in an unhappy marriage. Trim's answer, like Toby's whistling responses to his brother, is strictly non-verbal, but the reader has no difficulty in making sense of Tristram's ensuing comment. The twirl of Trim's stick, like the twirl in the capital letter 'J' of Jan van Eyck's signature in the middle of the Arnolfini double portrait, is an assertion of freedom and individuality – Trim's, Toby's, and, of course, Sterne's. For the reader has no difficulty either in grasping the message Sterne is conveying: the novelist's bondage to grammar and syntax is very similar to the bondage Toby envisages would be the result of his marriage to the Widow Wadman, and, even if, in Sterne's case, he has to submit to it most of the time, he will, like Panurge refusing the easy option of talking in French, go on asserting his essential freedom, his desire to twirl the stick of his fancy rather than conform to the conventions of novelistic narrative.

Celibacy means freedom, but it also means loneliness. Sterne's cavalier way with the conventions of novel-writing may mean the release of the spontaneous, but it also means that he is constantly skirting the abyss of arbitrariness or even meaninglessness. It is Toby's uncertainty, his wavering between alternatives, his lack of his brother's dogmatism and self-confidence, which saves him and keeps him human, just as it is Sterne's own awareness of the ease with which freedom can turn into meaninglessness which keeps his book alive and saves it from becoming a simple exhibition of cleverness. Toby's whistling, Trim's gesture with his stick, Sterne's introduction of black and marbled pages into his book – all these are poised at the junction of triumph and despair, and they remain moving and interesting precisely because they somehow signify both. Sterne's art, like that of Rabelais, requires exquisite touch precisely because both of them take seriously the fact that, with the demise of genre and of the boundaries established by tradition, art now has to make its own rules as it goes along.

22   Boundaries (2)

Over my notebook I sit hunched up. Over my typewriter, a little more upright. My hand moves over the page. My fingers hit the keys. I am writing.

But where is this 'I' who is writing? In my heart, which beats strongly as I work? In my head, where thoughts are swirling? In my anxious belly? My straining hand? My tapping fingers? Clearly it is in none of these places. And my sense of the absurdity of the enterprise whenever I try to write autobiography, to explain to myself why I am what I am and where I have come from and where I think I am going, proves to me that it does not reside in my memory either, in any continuity over time.

And yet I know too that those persuasive critics and theoreticians who talk about the death of the author, who argue that it is writing that writes and not 'I' – these I know are wrong as well. For when I am not at my desk, hunched over my notebook, a little more upright over my typewriter, it is I who suffer and no one else. And I know too that I can do something about it, can turn that suffering, that frustration, that confused desire, into a state at least as pleasurable as any I have experienced on the football field or in the swimming pool.

But how? Not, it seems, by sheer determination, any massive effort of the will. Not by the daily grind of training. That is the oddity of art, that it is very like sport in so many ways and yet not at all like sport in so many others. For, given a certain predisposition, certain physical attributes, faith in one's abilities and the capacity for hard work, anyone can play tennis or swim reasonably well. There is no equivalent to that in art.

Perhaps there once was. The painter and the composer learnt their craft from a master, as an apprentice, and had to fit their skills to the demands of patrons such as the Church or the nobility. All that has gone. Partly because the system of apprenticeship and patronage has gone, but that is only the symptom of a deeper change. It is no longer a question of fleshing out a given form. The form itself needs to be discovered or invented. And, once invented, needs to be reinvented with each new work.

The sense one gets, reading the letters of the great Romantic and post-Romantic artists – Keats, van Gogh, Kafka – is of enormous energy and desire finding it almost impossible to channel themselves adequately. The swimmer or runner can burn up energy in the pool or on the track and know that each day's work prepares him a little more for the race ahead; but, however much the artist would emulate him, he can't. For what would training mean in his case?

What is he to train *for*?

In retrospect of course one can often see a pattern in the career of a great artist. One can see that Proust's flurry of unfinished works between 1897 and 1907 was nothing but a groping towards *A la recherche*; that Picasso's whirlwind activity in those same years was leading him inexorably towards *Les Demoiselles d'Avignon*. But for Proust in those years it was a case of nothing but one failure after another, one confirmation after another that he was not a writer at all; and if Picasso's enormous talent and energy kept him from despair it could not hide from him the fact that he had still not found his real voice.

But how to find that voice, do what one feels one was set on earth to do? Does one, like Rilke, contain oneself in patience, waiting for the day when the spirit will condescend to visit? Or, like Kafka, go on pouring out sketches and fragments, work that refuses to add up, that does not and cannot satisfy, in the vain hope that suddenly one will see what has to be done and how? Or does one, like Coleridge, take the failure to fulfil oneself as the subject-matter of one's poem?

And what is satisfaction? Will it even be recognised when it comes? I know I have reached my goal if I win the race I have been training for, climb the peak I have set myself to climb. But goals in art are less tangible; often it seems unclear if they have been reached at all. Wordsworth and Coleridge, in their moments of dejection, remembered with nostalgia their childhoods and how different it had been then; Eliot, as he saw his work fragment under his hand, despite his best intentions, recalled, equally nostalgically, a culture that had once been unified and was now gone for ever; Kafka, in his later years, remembered with pleasure and despair the night of 22 September 1912, when, in one ten-hour stretch, he had written 'The Judgement' and felt it emerging from him as an organic whole, and felt 'how everything can be said, how for everything, for the strangest fancies, there waits a great fire in which they perish and rise up again'.

To be filled with the desire to make something, with the energy to work harder than ever before, but to be unable to discover what it is one wants to make, what kind of work one should be engaged in – that is the terrible frustration all artists are at one time or another a prey to – and have been since 1800. But it is of course not a frustration specific to the artist. It is a blight which fell on all emancipated westerners in the wake of the French Revolution and the demise of the Ancien Régime.* European literature in the nineteenth century reflects this malaise, littered as it is with works whose heroes dream of being Napoleon only to discover that they are petty murderers or frustrated clerks. For if anyone can become Napoleon the question then becomes: Why am I still only a clerk? With all that energy boiling up within me, with all that ambition to do something that will make the world sit up and pay attention, why can I not find the thing I need to do? The temptation dangled before so many heroes and heroines in nineteenth-century novels is the temptation of passion: a great all-consuming passion will finally release me from the pettiness of my surroundings, give meaning to my life. But, as Dante saw long ago, though passion may be better than indifference, it may in the end only be a mirage.

Describing in fiction the criminality or folly of a Raskolnikov, an Anna Karenina, an Emma Bovary, did of course bring some appeasement to their creators, just as describing his dejection brought appeasement to

---

* There are always counter-examples to frustrate the neat schemes devised by historians of ideas. Here such a one would be Diderot's *Le Neveu de Rameau*, written well before the Revolution but whose hero exhibits many of the symptoms I have ascribed to post-Revolutionary men.

Coleridge. But these are local and temporary palliatives, and we should not be surprised at the number of artists who have sought refuge in drink or drugs and, in the wake of addiction, often, in suicide. John Berryman summed it up in one of the finest of his Dream Songs, the 153rd, his response to the death of his friend, the writer Delmore Schwartz:

> I'm cross with god who has wrecked this generation.
> First he seized Ted, then Richard, Randall, and now
>   Delmore.
> In between he gorged on Sylvia Plath.
> That was a first rate haul. He left alive
> fools I could number like a kitchen knife
> but Lowell he did not touch.
>
> Somewhere the enterprise continues, not –
> yellow the sun lies on the baby's blouse –
> in Henry's staggered thought.
> I suppose the word would be, we must submit.
> *Later.*
> I hang, and I will not be part of it.
>
> A friend of Henry's contrasted God's career
> with Mozart's, leaving Henry with nothing to say
> but praise for a word so apt.
> We suffer on, a day, a day, a day.
> And never again can come, like a man slapped,
> news like this.

But of course it did, with the announcement of Berryman's own suicide a few years later. Which left no poet to mourn him with the plangency with which he had said farewell to his poet friends.

And yet, in spite of everything, that silent hunger *is* sometimes appeased. I no longer feel absent from the world or from myself. I no longer see myself sitting over my notebook or my typewriter. The pent-up energy finds an outlet.

How does this come about?

It comes, I think, from the discovery of boundaries. When I can reach out and touch an edge and it does not give, then I know that it is only a matter of time and patience before the entire boundary has been touched. And when there is a boundary there is a work.

Clearly I do not literally reach out my hand and touch something. But equally clearly this is not simply a metaphor. I do not imagine the boundary, I feel it. How?

Because I discover what happens when I cross it.

I do not know what the boundary is and I do not know where it is, but I know when I have crossed it. For on the other side there is nothing. No meaning and no foothold. Outside the boundary I am once more in the empty and arbitrary world from which I had thought to escape by starting to write.

The making of the work will then consist in my feeling my way forward innumerable times until this boundary, which I can only feel when I cross it, has been entirely mapped.

To discover the existence of a boundary is to discover the possibility of work, and of a work.

The work is that which lies inside the boundary. The work is the mapping of the boundary.

The temptation of addiction: to appease the frustration that comes from not being able to discover what work it is one should be engaged in.

The logic of addiction: to dissolve the boundaries between myself and the world so as to get rid of my *useless* isolation. But the boundaries do not disappear; they only recede as I advance.

The melancholy of addiction: if it is a kind of solution it is only a temporary and inadequate one, which takes us further and further from the world instead of returning us to it.

The dream of touch: that which joins me to the other by bringing both of us to life.

But it remains a dream so long as one is not working. So long as one is not advancing towards the boundary. Which means so long as one is not prepared to cross the boundary. To return to the empty room. The silent mirror. The temptation, the melancholy of addiction.

So long as one imagines that grasp, finality, can be substituted for touch.

The work I complete when I have brought the boundary into being is not constructed primarily out of words but out of gestures. A series of gestures which bring the words in their wake.

Touch

The structure consists of a series of gestures in a certain order which satisfies.

The structure is never final. As soon as it has been completed satisfactorily it ceases to matter. The search for boundaries begins again. It will always begin again. Not as Sisyphus rolls his stone up the hill again and again, but as the sun rises each morning, as one breathes in and out and then in again and out again.

Yet it is not as natural as breathing. Not even as natural as swimming or kicking a ball. For it is never possible to tell in advance where the boundaries will be or even if they exist.

There is no end to it. But ends no longer matter.

23　The Room (2)

It is very quiet in the room. The boy leans over the table, holding a card in his left hand. The operation is most delicate. His right arm, bent at the elbow, provides a solid fulcrum for his body; his left rests more lightly on the green baize, so that the card, when he will finally set it down, just so, on the fragile structure already erected, will not cause it instantly to collapse.

He is not tense, nor does he slouch. His head is held high, his neck firm, the black *tricorne*, fitting snugly on his head, shuts out whatever is above him. His large eyes, rather hooded, look down on the precarious structure before him.

Everything is still. The small drawer of the table is slightly open, jutting out towards us, but the boy is as unaware of this as he is of everything except his immediate task. Soon he will have placed the card he is holding on the fragile construction before him, and then either the whole edifice will collapse and he will have to start again from scratch, or, this step achieved, he will be ready to take up another card from the small pile in front of him and see what he can do with that.

At the moment, though, he has no thought for what is to come, his entire being is concentrated on the card he holds so lightly.

Four paintings by Jean-Baptiste Chardin.

8. (*right*) *The Young Draughtsman*, 1737. (Musée de Louvre. Photo © RMN)

9. (*below*) *Child with Spinning Top*, 1738. (Musée de Louvre. Photo © RMN)

10. (*facing page, top*) *Soap Bubbles*, 1738? (Metropolitan Museum of Art, New York; Wentworth Fund, 1949)

11. (*facing page, below*) *The House of Cards*, 1736–7. (National Gallery, London)

Now we are in another room. This time the boy has no hat but a fine wig fastened at the back with an elegant black ribbon. (Or perhaps it is another boy, slightly younger, perkier?) Instead of leaning on the table from the left and presenting us his right profile, he stands to the right of the table and we see him in three-quarters profile. The table too is smaller, and rather more crowded, with an inkwell in which rests a splendid feather quill, two books, a rolled-up parchment. These have been pushed to one side to make way for the top which the boy has clearly only just spun, though his hands now rest demurely on the table and his eyes watch the little top as though it had nothing to do with him. Empowered by the spin, the top has come quite close to the edge of the table and now hovers over another open drawer, though this one seems rather full. Striped red and green wallpaper at the back reinforces the sense of verticality imparted to the scene by the boy's upright stance and the fine quill, the dark inkstand and the top, which at present is just a little off the vertical.

In a moment it will fall over onto its side and come to a convulsive stop. But just now it is still spinning, almost still in its centripetal movement. No wonder the boy is gazing at it with such delight, a half-smile on his childish lips imparting to his face a freshness and naturalness quite at odds with the formal, elegant coat and shirt, the fine waistcoat with its prominent buttons, the powdered wig. But the hands tell us that this is a deft and agile child, not an overdressed mother's boy.

Now we are in a third room. Another boy, slightly older than the other two, equally elegantly dressed, the black *tricorne* on his head, but a long tress of hair hanging down his back this time, tied at the neck with a fine bow, leans on a simple table from left to right. His left elbow takes his weight and in his left hand he holds a pencil which he is sharpening with the knife held in his barely visible right hand. Covering most of the table is a large portfolio, tied by a ribbon, the ends of which hang over the edge of the table. He too is engaged on a task which, though requiring concentration, is mechanical enough to allow him to dream as he performs it, and his hooded eyes look down at his hands as though at a mystery at which he is content simply to be present.

And now we move out into the open. A young man leans over a stone parapet. Leafy branches surround him. A glass, four-fifths full of a milky liquid and with what might be a spoon sticking out of it, stands on the parapet beside him. He leans far over, anchored by his left arm which is

12. Jean-Baptiste Chardin, *Glass of Water and Coffee-pot*, 1761. (Carnegie Museum of Art, Pittsburgh; Howard A. Noble Collection, 66.12)

bent at the elbow, and resting his right arm on his left wrist. With the tips of the fingers of his right hand he holds a straw, one end of which is in his mouth. He is blowing bubbles, and a huge bubble has in fact just formed at the end of the straw, on a level with the parapet which is visible through the transparent film of the giant bubble.

His coat appears to be torn at the right shoulder, and his long hair, caught up at the back, has escaped at the sides, giving him something of the appearance of an orthodox Jew. To his left, a little surprisingly, a face is visible, topped by a curious twisted hat and cut off at the mouth by the parapet. Though the eyes of this figure appear to be staring straight at us, it is in fact at the bubble, which floats between him and us, that he is gazing. However, he is too much in shadow for us to be able to make this out with certainty. The eyes of the youth are lowered over his straw.

What these young men and boys are doing is in no way significant or important. They are not figures from Greek or Roman mythology, nor are they biblical or historical characters. There is nothing about them we need to know, for there is, in a sense, nothing about them that is worth knowing. The young man with the pen is probably a draughtsman, but what we see him engaged on is merely preparatory to the exercise of his profession. As for the other three, they seem merely to be passing the time.

Passing the time, but not only that, the scholars tell us: they are positively *wasting* time. These, we are told, are *vanitas* images, showing the
ways in which idle youths waste their time; or rather, the ways in which
Idle Youth Wastes Its Time. But the absurdity of this suggestion, which
has been current since his day, is so blatant that it merely highlights the
fact that there is something peculiarly resistant to interpretation in these
paintings of Chardin, so simple and so quiet, so wondrously beautiful. A
gap opens up between our immediate, physical response to them and the
ability of the intellect to provide explanations for them: Why these boys?
Why doing these things? What is Chardin trying to tell us?

Proust, who did not write about these works but did sketch out a little
essay on Chardin's still lifes which he never completed, is, as one might
expect from someone who knew how to look and could think about what
he saw and then express it powerfully and elegantly, closer to the mark.
He suggests that the still lifes, those beautifully painted arrays of pots
and jugs and simple kitchen shelves, give us the feeling of the quotidian,
not in the sense that they are ordinary but in the sense that they are felt
to be in daily use. The very way the paint is put on conveys to us not only
the constant use to which the depicted objects are put but also the loving
care with which they are made. The worn roughness of their surface, so
different from the high polish of most contemporary still lifes painted in
order to demonstrate the status and wealth of their owners, conveys at
once the quality of such craftsmanship and the wear and tear they have
endured in years of household use. And this becomes the central feature
of the painting itself, as though Chardin's own work on it was akin both
to that of the craftsman who made the utensils and of the maidservant
who uses them day in day out as she goes about her work.

Proust, who knew all about the iterative mode, all about the skill
required of the novelist if he is to convey the sense of the regularity of
days which pass by, each essentially the same, rather than the exciting
plot development which is the staple of the common novelist, could well
appreciate what Chardin was up to. At the same time he could bring out
what Chardin's contemporaries and later commentators have found so
disturbing about his canvases: the complete absence of narrative. One
has only to compare these still lifes with those of other artists (Morandi
is of course the great exception, the one painter who seems to have taken
Chardin's lesson to heart) to see that while they are all trying to make a
point, either about their own skill or about the way we live or should live,

he is concerned only with the reciprocal gifts he and those pots and pans bestow upon each other: they, through the fact of their use and continuity in the ordinary and the mundane, give him the strength to do his own ordinary and repetitious work, while he, through the power of his vision and the skill of his hand, is able to celebrate them for what they are.

It is as though Chardin were telling us that everything is in danger of turning into an anecdote or a lesson, and therefore in danger of ceasing to be itself. It requires a positive effort to take a stand against the steady erosion of mundane reality, an effort all the more problematic in that there seems to be nothing that underpins it and lends it authority. For the authority has to come from the work itself – the work of the hand and the finished work of art. It takes a very special kind of artist to do what Chardin does here.

In the four paintings in which figures appear the refusal to tell a story is even more striking because it seems to go so much against the grain of figurative art. But the point of catching these youths in such intense yet unguarded moments is that for once in the art of the West our viewing time and the time within the painting coincide: we stand there, entranced, and we know that as we turn away the hand will come forward, the card will be put down, the top will fall over, the bubble will burst. But so long as we look, time is suspended, everything waits. Or rather, time is gathered up and everything stands in a state of dynamic repose, concentration and relaxation together.

'It is said that he has a technique all his own', Diderot wrote about Chardin, 'and that he uses his thumb as much as his brush. I don't know if this is true; what is certain is that I have never known anyone who has seen him at work.'

No one may have seen Chardin at work, but all his paintings tell us how he worked: slowly, patiently, modestly, intensely, taking pride in his craft but with no very exalted notion about the status of the work of art or of his own position in any hierarchy. Skill is secondary, he is said to have remarked, what is important is empathy with the subject. History-painting did not interest him. Nor mythology. His work is at once disturbing and exhilarating because he shows us that there is no mystery about art, but that it takes a lifetime to master it; no mystery about life, but that it needs to be lived, not endured. His work, like that of Morandi, recalls us to art and to ourselves. As Ponge, another artist who, like Proust, knew something about the strangeness of the ordinary, put it:

When the ancient mythologies are no longer meaningful, *felix culpa*, we begin to experience the humdrum reality in a religious mode. I think we will be more and more grateful to those artists who, by their silence, by their simple abstention from the themes imposed by contemporary ideologies, have shown their solidarity with the non-artists of their day [auront fait preuve . . . d'une bonne communion avec les non-artistes de leur temps].

And he adds, lest we miss what there is of the awesome in Chardin:

Chardin maintains a laudable balance between the tranquil and the fateful. For my part, the fateful is all the more obvious in that it advances at a leisurely pace, without showy outbursts, taken for granted. This is 'sanity'. This is our beauty. When everything comes together naturally, in a pre-ordained lightning. [Le fatal, quant à moi, m'est d'autant plus sensible qu'il va d'un pas égal, sans éclats démonstratifs, va de soi. Voilà donc la 'santé'. Voilà notre beauté. Quand tout se réordonne, sans endimanchement, dans un éclairage de destin.]

Those quiet rooms. Those strange, still youths. They are unimaginable and unimagined – till he painted them. Here too vision and skill go hand in hand. We can talk of Chardin's extraordinary colour sense, of the brilliance of his modulation, of the cunning way he realises space, of the evenness with which he paints the faces of his protagonists, the tables, knives, glasses of water, cards and green baize; we can note the skill with which the snail-shell spiral is made to run from the tip of the *tricorne* down through the body and out from the pen, the card, the quill. But, as Chardin said, this skill is secondary and what is important is the empathy with the subject. Something is happening that is both trivial, infinitely repeatable, and momentous, unique, unrepeatable. *This* bubble, once it has burst, will never exist again; *this* particular house of cards, once it has collapsed, will be gone for ever; *this* particular trajectory of the top, once the top has fallen over, will never be repeated; *this* particular pencil will never be as long (and the boy sharpening it will never be as young) again.

The eyes look down and outward, at the hand which is sharpening the pencil or laying down the card, at the spinning top or bubble which has just been released; but the gaze is, somehow, emanating from the whole body, just as it is the whole body which is focused on the act of sharpen-

138

ing the pencil, spinning the top, building the house of cards. That is what is meant by absorption. And our gaze too, as it moves from the hooded eyes to the hand holding the card, the pencil, the straw, falls under the same spell. We too are in that quiet room, not thinking anything, not doing anything purposive, anything that 'needs doing'. This is order. This is freedom. The hand releases the card as we release the painting that is before us, as *Oedipus at Colonus* released its protagonist, accepting that we must let it go and yet, for a moment, holding it within the orbit of our attention.

Appendix

'I touch you' – but 'I am touched', 'that is touching', mean something quite different. 'I am touched by this branch' can only mean: this branch moves me (because of its leafiness, bareness, etc.). If we mean it literally we say: 'This branch is touching me.'

'I am touched.' What gives rise to this apparently figurative expression? The notion of the *heart* being touched? A hard-hearted person is one whose heart cannot be touched. Is the concept of the soft heart at the back of the phrase 'a soft touch' – someone who can easily be persuaded to part with his money? (cf. 'He touched me for a fiver.')

'I am touched' means: I am moved. But 'he is touched', now an archaic expression, means that he is not quite normal. Touched by what then? Fate? God? Ill-luck? (But of Dr Johnson it could have been said: 'He was touched' – by Queen Anne, for the scrofula.)

Stuart Hood, in a letter, reminds me of the custom Italian males have of touching their testicles to ward off the evil eye. 'I remember an Italian friend of mine saying', he writes, 'that at the *liceo* they would go into the exam room *ter quaterque testiculis tactis.*'

'I am tactful', 'he is tactful', etc. How strange to be full of touch. The Shorter OED gives, under 'tact' 1.b.: '*fig.* A keen faculty of perception or discrimination likened to the sense of touch'. Under 2: 'Ready and delicate sense of what is fitting and proper in dealing with others, so as to avoid giving offence or to win good will'. Which begs the question of why a keen faculty of perception or discrimination *should* be likened to the sense of touch.

'Tactile', on the other hand, remains firmly unmetaphorical. (And 'tactic' comes from a different root, the Greek *taktike*, related to *tekne*, from which our *technique* derives.)

'Toccata'. 'A composition for a keyboard instrument, intended to exhibit the touch and technique of the performer, and having an air of an improvisation'. Thus the Shorter OED. But the New Zingarelli asserts that it is 'Atto del toccare una sola volta', the act of touching once only, and gives the musical definition: 'Sonata di un sol tempo, di stile elevato e da eseguirsi uno strumento a tastiera' (a keyboard instrument). Stuart Hood again: 'Clearly a feminine past participle used actively. But what is the feminine noun that is to be understood? Presumably the same one as in *sonata*, which is also a past participle. Can it be *musica*?'

On the keyboard a single touch will produce a single note. Is there the indication in the definitions of 'toccata' that this will reveal one's strengths and weaknesses more clearly than a stringed instrument or one that has to be blown into.

'What a fine touch he has!' Of a pianist of course, but also of a footballer or tennis player. But what of the diver or gymnast? It seems that something else must be touched and then released, such as a keyboard or a ball. But would the term be used in snooker, where the essence of the game is to have a fine touch?

On the other hand, while 'I have lost my touch' could be said by a pianist or a centre-forward, it might also be said by a teacher or a theatre director. Note the difference here between 'I seem to be losing my touch' and 'I seem to be losing touch with them.'

'A nice touch.' The icing on the cake.

'A light touch' – but 'a heavy hand'. Is that because touch cannot by definition be heavy? But then is not 'a light touch' a tautology? Well, not *exactly*.

'How touching it was to see them' – but 'she was so touching' is more likely to be used of someone we see acting a part: 'She was so touching as Juliet.' 'She was so touching at her father's funeral' would be either silly or catty. Thus we recognise that our hearts can be touched ('I was so touched that she remembered'), but we may suspect the person who touches us in this way of having designs upon us.

'*Touché*!' The cry of the fencer – nowadays, naturally, only metaphorical. The admission of the one who has engaged us in an intellectual fencing match that he has been fairly beaten. Yet, as is the way with metaphors, it can be a means of warding off a real admission of defeat: *Touché*! – but have I really been touched by your argument?

'Stay in touch', we say, and are hardly sure whether we mean it or not.

Notes

p.6. Merleau-Ponty – Maurice Merleau-Ponty, *La Prose du monde*, Gallimard, 1969, 186–70; *The Prose of the World*, tr. John O'Neill, Heinemann, 1974, 134–7.

pp.9–10. 'Photography maintains' – Stanley Cavell, *The World Viewed: Reflections on the Ontology of Film*, enlarged edn, Harvard University Press, 1979, 23, 26.

p.10. Walter Benjamin – 'The Work of Art in the Age of Mechanical Reproduction', in *Illuminations*, ed. Hannah Arendt, tr. Harry Zohn, Jonathan Cape, 1970, 224–5.

p.12. Donald Davie – *Articulate Energy*, Penguin Books, 1992 edn, 259–60.

p.15. 'I feel I was wrong' – Marcel Proust, 'Journées de pèlerinage', in *Contre Sainte-Beuve*, ed. Pierre Clarac et Yves Sandre, Gallimard, Pléiade, 1971, 84–6; *On Reading Ruskin*, trans. and ed. Jean Autret, William Burford and Phillip J. Wolfe, Yale University Press, 1987, 16–17.

p.20. 'Make all men living' – Aeschylus, *Choephoroe*, 1. 902. See John Jones, *On Aristotle and Greek Tragedy*, Chatto & Windus, 1962, 100–103.

p.20. Ezekiel 33 – see Harold Fisch, *Poetry With a Purpose: Biblical Poetics and Interpretation*, Indiana University Press, 1988, Ch.4.

p.25. Foucault has written eloquently – *Les Mots et les choses*, Gallimard, 1966, Ch. 1.

p.26. too low in realistic terms – see Craig Harbison, *Jan van Eyck: The Play of Realism*, Reaktion Books, 1991, 34.

p.26. Some have suggested – see, for example, Jill Dunkerton, Susan Foister, Dillian Gordon, Nicholas Penny, *Giotto to Dürer: Early Renaissance Painting in the National Gallery*, Yale University Press in Association with National Gallery Publications Limited, 1991, 260.

p.30. an episode – *A la recherche du temps perdu*, ed. Pierre Clarac et André Ferré, Gallimard, Pléiade, 1954, I. 27–43.

p.33. 'identical emotions' – *ibid.*, 155; *In Search of Lost Time*, tr. C. K.

Scott Moncrieff and Terence Kilmartin, revised by D. J. Enright, Chatto & Windus, 1992, I. 186.

p.33. '[s]ometimes to the' – *ibid.*, 157; 187.

p.35. 'Today the room' – Georges Perec, *La Vie mode d'emploi*, Hachette, 1978, 331; *Life: A User's Manual*, tr. David Bellos, Collins Harvill, 1987, 261.

p.37. Auden – W. H. Auden, *New Year Letter*, Faber and Faber, 1941, 121.

p.38. 'sanza speme' – Dante, *Inferno*, IV. 42. I have used C. S. Singleton's edition and translation, *The Divine Comedy*, Princeton University Press, 1970.

p.41. Jill Mann – in an unpublished lecture on *Inferno* V which she was kind enough to show me

p.42. William Burroughs – *Junkie*, Penguin edn, 1977.

pp.42–3. Stanley Cavell – *op. cit.*, 102.

p.44. Graham Greene's comments – 'The Lost Childhood', in *Collected Essays*, Penguin Books, 1970, 13.

p.46. 'And it was then' – Fyodor Dostoevsky, *The Devils*, tr. David Magarshack, Penguin Books, 1953, 63.

p.48. 'Do you know' – Proust, *op. cit.*, I. 163; I. 195.

p.51. 'Tell me' – I have used E. F. Watling's Penguin translation throughout: *The Theban Plays*, Penguin Books, 1947.

pp.59–60. St John – I have used the King James Bible throughout.

p.60. Peter Brown – *The Cult of the Saints: Its Rise and Function in Latin Christianity*, University of Chicago Press, 1981, 88.

p.64. 'heals the sick' and 'He touches' – Marc Bloch, *The Royal Touch: Sacred Monarchy and Scrofula in England and France*, tr. J. E. Anderson, Routledge & Kegan Paul, 1973, 81, 83.

p.64. *darshan* – E. A. Morinis, *Pilgrimage in the Hindu Tradition: A Case Study of West Bengal*, Oxford University Press, 1984, 73. My thanks to Stephen Medcalf for introducing me to this book.

p.65. 'Peter Brown' – *op. cit.*, 87.

p.66. 'the distance' – *ibid.*

pp.66–7. 'The image' – Morinis, *op. cit.*, 182.

p.67. 'In Tebessa' – Brown, *op. cit.*, 87–8.

p.69. Berkeley – *A New Theory of Vision*, XLI. I am grateful to Bernard Harrison for directing me to Berkeley's essay.

p.70. 'How many times' – Proust, 'Journées de lecture', in *Contre Sainte-Beuve*, 194; *On Reading Ruskin*, 128–9.

p.73. 'Benares is to the East' – Morinis, *op. cit.*, 74.

p.73. Craig Harbison – *op. cit.*, 187.

p.73. Eamon Duffy – *The Stripping of the Altars: Traditional Religion in England 1400–1580*, Yale University Press, 1992, 199.

p.74. 'You knight of Christ' – quoted in *ibid.*, 205.

p.74. A fascinating Lollard text – 'The Testimony of William Thorpe 1407', in *Two Wycliffite Texts*, ed. Anne Hudson, Oxford University Press, 1993. The quotes come from lines 1229–1389. I have modernised the spelling.

p.76. 'for as moche' – quoted in Duffy, *op. cit.*, 580.

p.76. 'By the end' – *ibid.*, 582.

p.77. It has often been remarked – the classic work is Louis Martz, *The Poetry of Meditation*, Yale University Press, 1954.

p.79. 'A hectic trade' – Brown, *op. cit.*, 88–90.

p.79. 'His body lay hidden' – *ibid.*, 91–3.

p.80. 'the invisible gesture' – *ibid.*, 97.

p.80. 'While the relic' – *ibid.*, 100–101.

p.81. 'shall not show' – this and the following quotes in Duffy, *op. cit.*, 584–5.

p.82. 'The price for such accommodation' – *ibid.*, 593.

p.85. J. A. Burrow: *A Reading of Sir Gawain and the Green Knight*, Routledge & Kegan Paul, 1965.

p.86. Father Zosima – Fyodor Dostoevsky, *The Brothers Karamazov*, tr. David Magarshack, Penguin Books, 1958, I.46.

p.89. Georg Christoph Stirm – Stephen Greenblatt first drew attention to this letter and published it in 'A Passing Marvelous Thing', *Times Literary Supplement*, 3 Jan., 1992, 14–15.

p.91. Nicholas Thomas – 'Licensed Curiosity: Cook's Pacific Voyages', in *The Cultures of Collecting*, ed. John Elsner and Roger Cardinal, Reaktion Books, 1994, 134.

p.92. Krzystof Pomian – *Collectors and Curiosities: Paris and Venice, 1500–1800*, Cambridge, 1990.

p.92. Humanist popes – see Donald Horne, *The Great Museum: The Representation of History*, Pluto Press, 1984, Ch.1.

p.95. 'It has been estimated' – Robert Harris, *Selling Hitler: The Story of the Hitler Diaries*, Faber and Faber, 1986, 183–4. The other quotes in this paragraph come from pages 184–7.

p.95. 'included an almost complete set' – *ibid.*, 111.

p.96. a specific genre of post-war art – see Saul Friedlander, *Reflections on Nazism: An Essay on Kitsch and Death*, Harper and Row, 1984.

p.96. 'Why should anyone pay' – Harris, *op. cit.*, 387.

p.102. the Age of Suspicion – see Nathalie Sarraute's famous essay, 'L'Ère du soupçon', in *L'Ère du soupçon: Essais sur le roman*, Gallimard, 1956.

p.111. 'Since his illness' – Jonathan Cole, *Pride and a Daily Marathon*, Duckworth, 1991, 148.

p.111. 'Nothing is granted to me' – Franz Kafka, *Letters to Milena*, ed. Willy Haas, tr. T. and J. Stern, Schocken Books, 1953, 219.

p.127. 'how everything can be said' – Franz Kafka, *Diaries*, ed. Max Brod, Penguin Books, 1972, entry for 23.9.1912.

p.136. Proust – 'Chardin et Rembrandt', in *Contre Sainte-Beuve*, 372–82.

p.137. 'It is said' – Denis Diderot, 'Le Salon de 1767', in *Salons*, ed. Jean Seznec and Jean Adhémar, Oxford University Press, 1983, III; there is an English translation by John Goodman, *Diderot on Art*, Yale University Press, 1995, II. 86.

p.138. 'When the ancient mythologies' – Francis Ponge, *Nouveau receuil*, Gallimard, 1976, 171–3.

Index

Index

Index